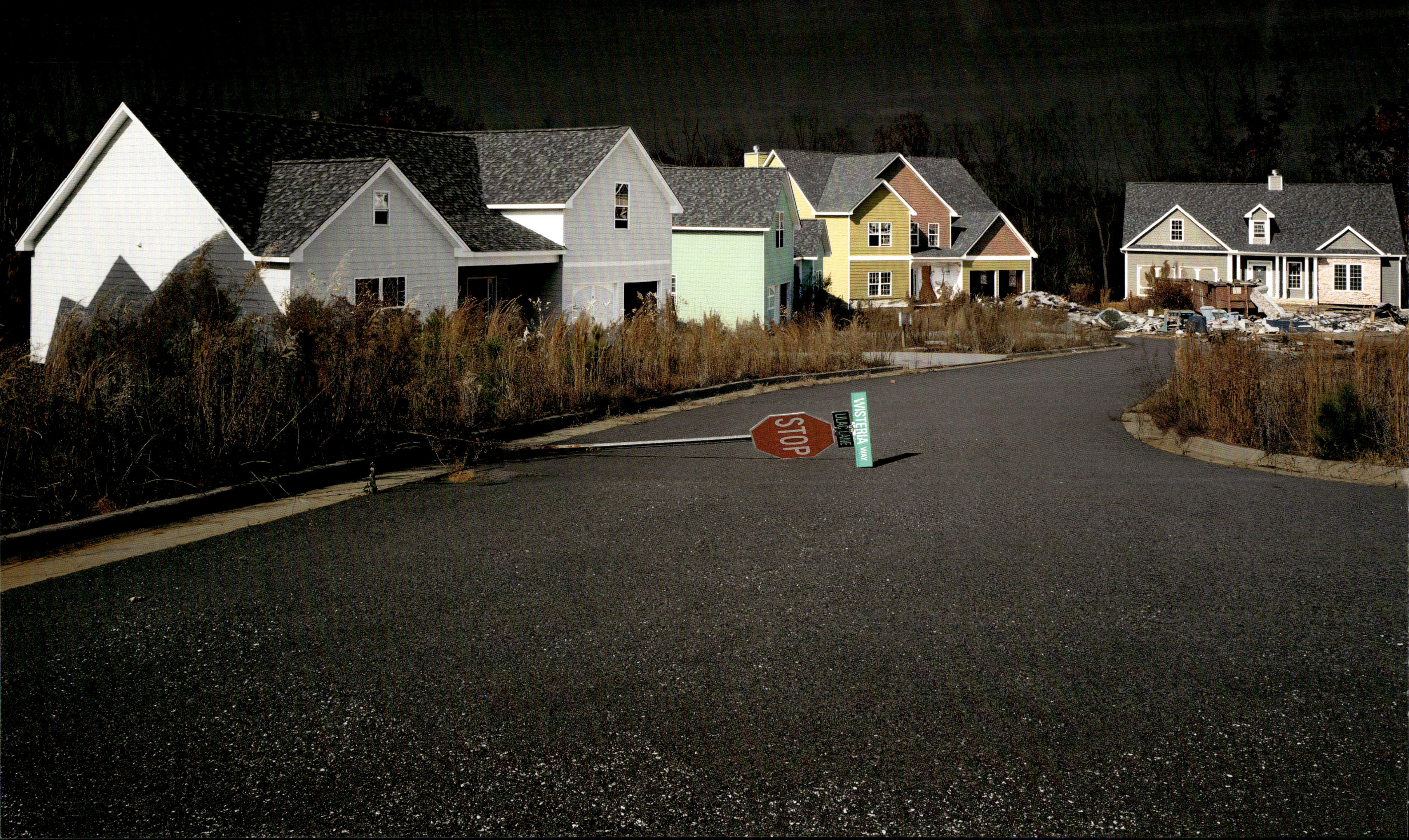
STOP
LILAC LANE
WISTERIA WAY

EDGAR MARTINS THIS IS NOT A HOUSE

dewi lewis publishing

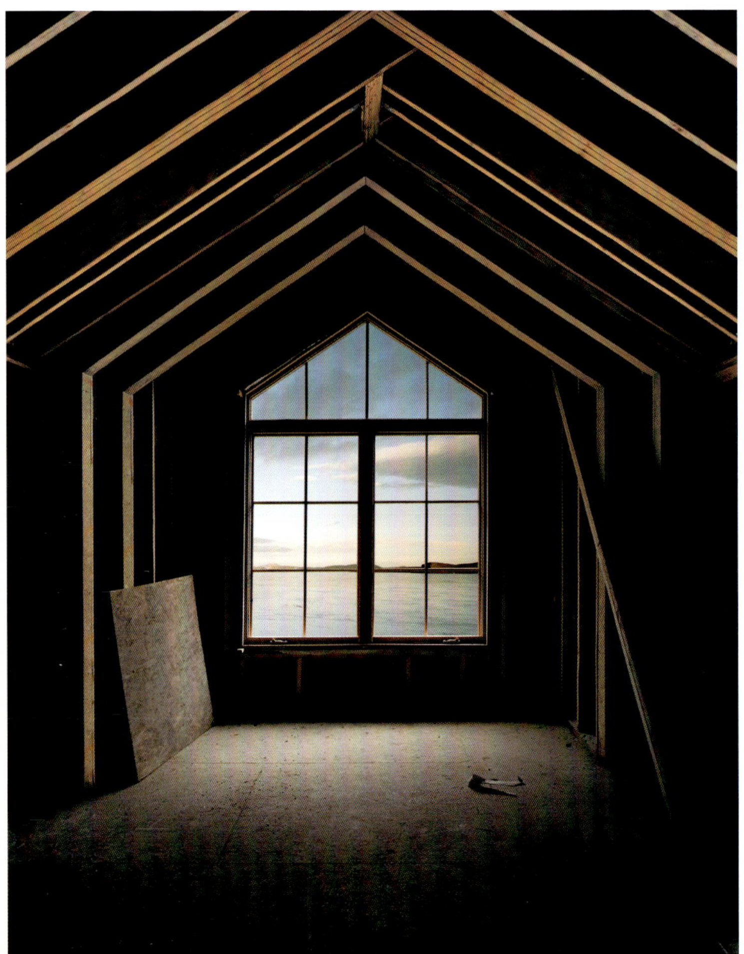

SACHA CRADDOCK UNDER CONSTRUCTION

The relationship between institution and artist somehow guarantees that art can function, that it is part of life rather than separate. All artists work to commission; in fact, it is now difficult for even a painter to do anything other than work on a series to order. It is almost impossible for an unestablished and not privately funded artist to work speculatively. A commission establishes the relation between context and place, between need and finance: a sculpture, sound piece, performance, book, mural, or photo series for a wall, a book, a courtyard or even for an online or print newspaper. When the artist is a photographer, however, the question of role can get more difficult.

The case of Edgar Martins and his photo essay for *The New York Times Magazine* is clear, but fascinating. It started well; there is a crisis in the housing industry, so let's ask an artist to show us – the paper, and the public – what a newspaper photographer never could. After all, is not an artist able to show us what we cannot see? The role of the artist, post Romanticism, must be to help us approach things from a different angle, to cast light upon something that might not be found any other way. This is the dominant notion of art and its role in society; the excuse for art too, because, perhaps for some, it still needs an excuse. Art is to serve society; it has to fulfil a role, unlike in the simplistically characterised bad old days of 'art for art's sake'. But, like the 1970s boyfriend who, on finding his girlfriend to be sentient and complicated, says he finds the contents of the parcel different, *The New York Times*, which prides itself on accuracy, withdrew Martins' work from its website on hearing that the images had in fact been 'manipulated'.

Photography is still expected to serve, though, and the art photographer is forever trying to separate, to break away from this notion of function. No art photographer wishes solely to represent. Of course, everything has to be a matter of choice, of surprise even, and to play with clarity, with an exact image, further extends the open, three-dimensional power of art to come from any angle. The breakdown of formal distinctions is now so complete that the photographer, who seldom any longer calls his or herself just that, not only takes photographs but is also an artist in other media. The sculptor, on the other hand, who would not dream of calling him or herself anything other than an artist, is just as likely to consider the photograph as 'found'.

Yet is Edgar Martins still seen as a photographer in that he only makes photographs and books of photographs? From work made at the Royal College of Art onwards, his photographs of empty, strange, heightened unreal places in contrasting light have been exceptionally strong. The role played by Martins' work is complex, however; the picture itself is a mere beginning of an extension of physical and mental possibility; the atmosphere remains consistent throughout, and landscape and urban landscape are treated in the same way. These images, which represent the beginning of an idea, are a mere means to a flow of associative thought that goes far beyond observational truth into individual and collective understanding. Fantasy and theatre, face value observation, surreal and unreal atmosphere, are united beneath a surface of production.

So Martins was invited to make a photo essay on, about, the collapse of the housing market in the United States. By asking Martins, *The New York Times Magazine* was obviously expecting imagery that would go further than anything a newspaper photographer could produce. Even the most 'unusual' winning photograph in a World Press award could not have the same range of quality as a work by Martins. For a start, the touching, formally grave use of light and dark, the volume of mass, the invention as well as the knowledge of art history, already recognisable in his previous work, have no part in reportage.

Reasons for making a picture, or taking a picture, are complicated at the best of times, and Martins' self-motivation guarantees that his work has never been illustrative of anything other than, perhaps, his own invention. The common thread running through the works, however, is a formal quality, where the small shifts achieve an extension of an historical commonwealth of language. Neither reportage, nor documentary, the work exists, in part, for another world.

'The Ruins at Holyrood Chapel' at the Walker Art Gallery, Liverpool, was painted as an extension of Louis Jacques M. Daguerre's play with illusion. This very romantic view of the Chapel ruins, by the father of photography, with heightened, glowing moonlight seen through a filigree of partially collapsed Gothic window frame, is almost monochromatic. The painter here makes a sort of prop for a commonly established and anticipated visual language. The ruin, from Dürer's 'Madonna with Iris', with Mary in front of a set of scaffolding, throughout visual and literary language, indulges a sense of romantic past, of irrational transience; it also acts as a metaphor for the value of art over and above the representation of fact.

Of course, the buildings in Martins' series are all, in any case, ready-made ruins in that they are already in a state of flux, some way between disintegration and construction. His awareness of the history of the ruin in painting and photography allows the artist to step into this to provide an instant metaphor for non-place, non-existence, and the ultimate value of the manipulation of expectation.

Daguerre, inventor of the daguerreotype, used the stage to set the scene, to adopt a theatrical approach right there at the very beginning of the history of photography. So much later, after the medium has fully merged with, and then overtaken, painting as the provider of portraits of soldiers, sailors, street vendors, as well as landscapes, battle scenes, state openings, and criminal and anthropological records, the digital era ends the relationship between captured moment and truth. Photography, with the use of post-production, can reveal itself as the unfocussed harbinger of broken time, to set up and aid the representation of reality.

The ruin here does not refer only to American photography and past coverage of downturn and depression. Rather, it is rooted in the history of painting. A post-modern tendency to equate the representation of something almost with the thing itself can lead to terrible problems. The photographic commission is a difficult thing to agree to or want, because it lends itself to description, to the artist doing what is, in a way, expected. When the editors of *The New York Times* asked Edgar Martins to deal with, in the artist's words, the 'shame' of these partially built, abandoned, usually aspiring luxurious places, for both their magazine and the accompanying slide show on the *NYTimes.com* website, they must, in a way, have wanted him to say something expected, but different perhaps. But where was the art to 'lie'?

The Jean Nouvel building in New York; empty, unfinished, with flapping plastic and touching graffiti, presents an edge of fashionability with orange plastic, a series of planes, ready for an artist to see, or capture, or make happen in his or her terms. But how to approach the subject – go into it, at it? And what is that subject? A commission for a newspaper seems to involve a different set of considerations; truth to story, lack of embellishment, a level or approach that implies a true and honest take on the subject. Journalistic words are gathered by a combination of hunch, habit, received knowledge, politics, expectation, and stance. A subject, such as the downturn and virtual halt of the housing industry, is a given, the idea anyway dripping with strange pathos, about something that was meant to happen but never materialised in a more general context of profligacy, waste, breakdown and poverty.

Martins had submitted a synopsis to *The New York Times* that followed formal and factual lines on the new experience of architecture. He called it 'Ruins of the Second Gilded Age' (after the title of the feature). But a subject like this is also no subject; it can be really individual in terms of experience, disappointment and bankruptcy, and equally general, open, factual, fictitious, and emotional. How can this best be represented? Why would, or could, an artist, who happens to be a photographer, make something different, and how can the photograph manage to be a source of surprise? Was it Martins' job to transcend and extend the point at the same time? Does a commission inevitably carry with it a sense of disappointment? How can a subject be bettered? There has to be a way to speak differently, for an artist not to change a subject, yet 'say' or 'express' it better?

The places photographed by Martins were not dead but alive, unravelling. Perfection has not been achieved; the places, hidden and shaken, have been torn apart by visitors taking things away. Everything that can be taken has been removed; sections of floor cut to order, to fit housing elsewhere. The Show Home becomes a site of underdone undoing, never having reached that pinnacle of high finish; the shiny, fresh, promise of a new, albeit fake, old. So the real changes very fast and who is to say what is what, when wind, weather, human need, and greed, knock at a door that is about to be lifted from its own hinges. Is not the composition of a picture, the arrangement, as much a matter of time as anything else? Walker Evans, an important documenter of a different American downturn, would perhaps have arranged the furniture in a room before photographing the scene.

This commission provided Edgar Martins with a classically fictional, staged opportunity as he photographed vacant stages, portentous but still unvisited by the normality they promise. The empty shell of the stage prop home, in various stages of completion, promotes a pretence and projects a fiction; a serving suggestion to be served up by an artist intercepting an idea. A serving suggestion already set up anyway, from hunting lodge to high modern, with its inbuilt range of thematisation. From the pile of rubble fed through the side of an otherwise virgin building. From the leaves on the floor but also from the extreme shift between the artist's preferred relationship to his audience, through the production of large vertiginous prints which release illusory space free of gravity, to the very different context of a newspaper magazine and online slide show.

Except nobody can stay in the real space of these gated separate communities which will, in this case, never materialise. There is perhaps only one person in all these pictures. The only participants are actual members of an audience in front of the huge reproductions, which tip, or tilt, against the forced march of architectural symmetry. The illusion places the spectator there, really there, but it remains unreal as the artist and audience are expected to play judge, jury, defendant and even custodian of a general idea. The overview is indeed achieved, the artist doing what he can. Whole streets, infrastructures, new roads, corners, sign posts, waiting, half-wrecked, appear like a false town centre erected solely for soldiers training in contemporary urban warfare.

Martins explains that he uses symmetry to reinforce the image of architecture; he places us in the undoing of a subject as much as its construction. His work suggests a relation to the architect's 'artist's impression'. Architecture is purely notional, from proposal, through drawing and three-dimensional simulation, right up to and even including the building process itself. Long before computer graphics, it has been important to project, draw, collage, and model how something might eventually look; with streets lined by trees, people for scale, and children running down the projected sloping bank. Here in the case of *This Is Not a House*, everything is also pared down, laid bare, naked and fast becoming a ruin. Martins takes time backwards and forwards with this symmetry. The new is lost, prematurely aged, in depictions of collapse and breakdown; balanced, literally, by the return of these uninhabited areas to the bare bones of possibility.

The image itself is built up of many parts, many components. The mirroring of architectural elements conveys this. Martins makes symmetry almost fearful in its insistence. An image of Dawsonville, Ga, a suburb of Atlanta, 'was conceived by making use of multiple flashes'. This build-up of elements not only reflects the proposal stage of an architectural cycle but also alludes to the moral understanding of a language that matches ambition that will collapse. Perhaps the structures have not all been conceived by architects, and are just as likely well-worn plans laid down by builders in a mock local genre. The mirroring from side to side enters the point where the idea matches a reality that marches backwards. But so much of this lends itself to a discussion of time. Is this as the artist found it when he arrived? Is he being true to the idea or moment? This is not a fast-changing situation involving the actions of others. Did the commission afford, allow, a real relationship with place? Were the visits fleeting?

How moralistic in terms of individual experience, how keen the insistence on physical interaction, on the authority of individual sight at the moment of seeing. Not exactly static, these 'ruins', which are in a perpetual state of undoing, are in retreat from expected use. Things do not change as if hit by a bomb here, and death is offstage. Misfortune is not made suddenly visible as in any natural disaster. Breakdown here is not an act of God: it happens. Going backwards in time, unravelling labour, re-tracing effort, the financial confidence that financed the construction was, in any case, a chimera.

Places often feel on the edge of somewhere, and the pictures cannot but help to allude to the loneliness of the expanded town, the hinterland of half-light, of waiting at the edge of a nature keen to encroach once again. The solitary wait on the edge of expansion reminds us of Edward Hopper's paintings of the colonial dream threatened by loneliness. The property slump, the housing crisis in other words, is not exactly natural, but capitalistic growth and collapse also has a cyclical nature.

There are no figures though, because none are there. Martins enters a world of metaphor so easily: it is there, that unreality, and you could never make it up. He need never make it up because it is there to see, to know, but not to grasp. A photo essay takes the general and yet the artist is able to grasp the strange, the fictitious, the uncanny, the uncontrolled; to display a desire for romantic fulfilment in artistic rather than personal terms, for the projection of self into another place, for camping it seems, precariously, in a space on the way out of possibility.

The story starts with a sense of expectation but soon runs into notions of honesty and dishonesty, truth and lies in photography. Round and round like an old record this plays, never going away. An artist might be thoroughly versed in the fact that image is pure construction, in part because it is the artist who constructs the image, but the newspaper editor, who also constructs the news, might have difficulty understanding the nature of that construction. The role of the photographer gets mixed up, in this particular case, with a sense of truth to story more normally expected of a journalist. The artist loses out to the notion of the photographer as conveyor of truth. However manipulative they might be, a particular combination of words can be broken down and challenged, while the speed with which the overall effect of an image arrives at the eye means it functions in a very different way.

THIS IS NOT A HOUSE

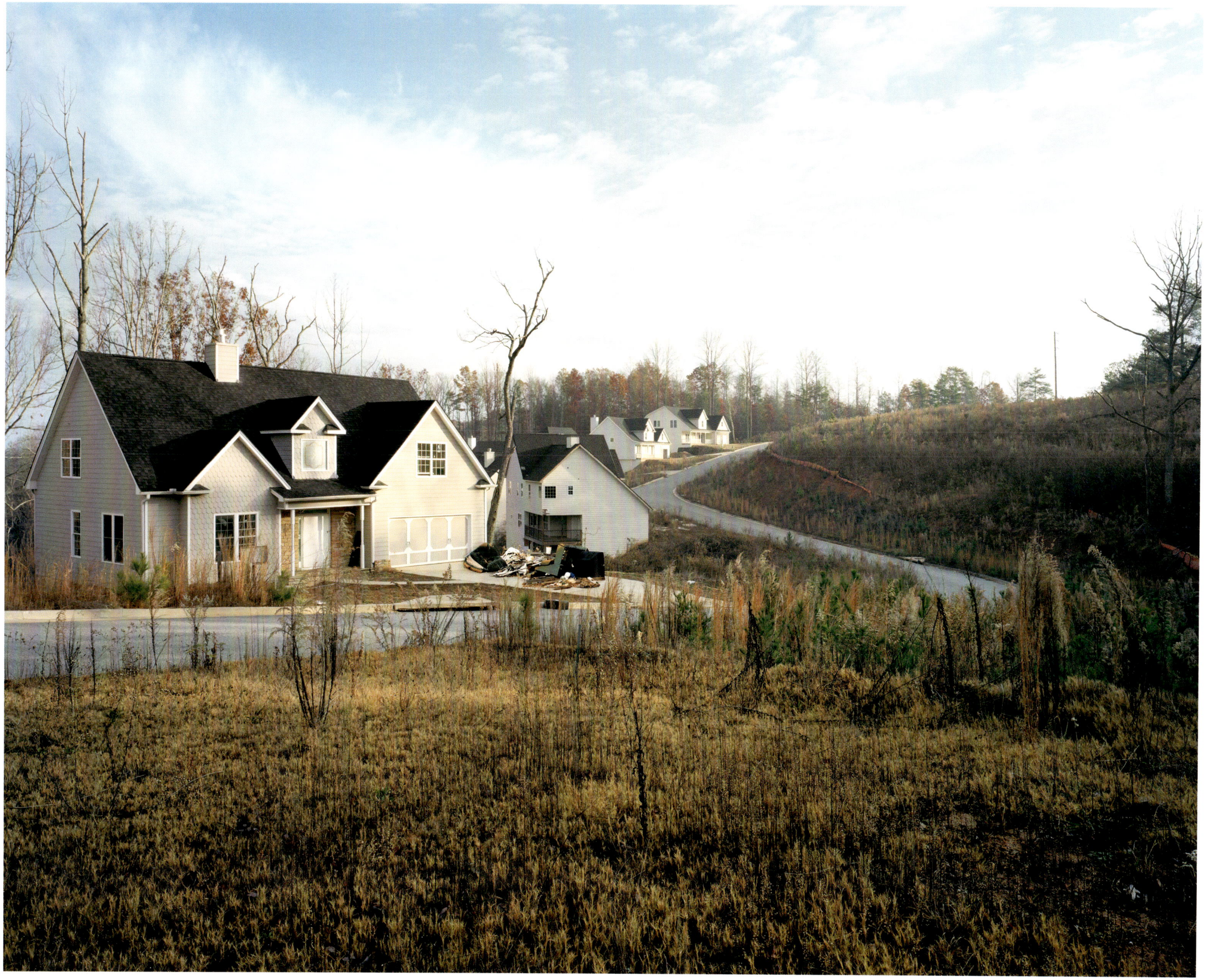

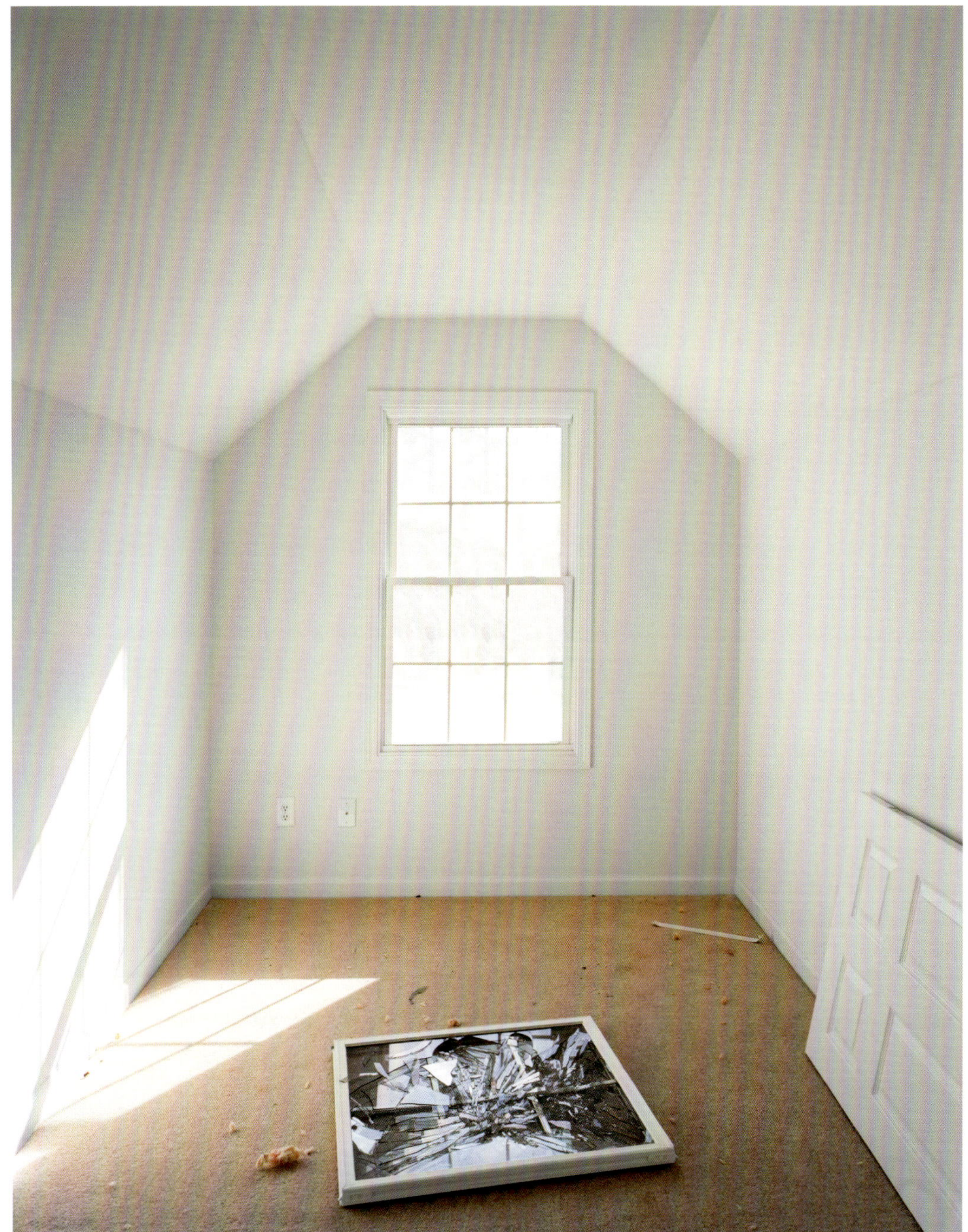

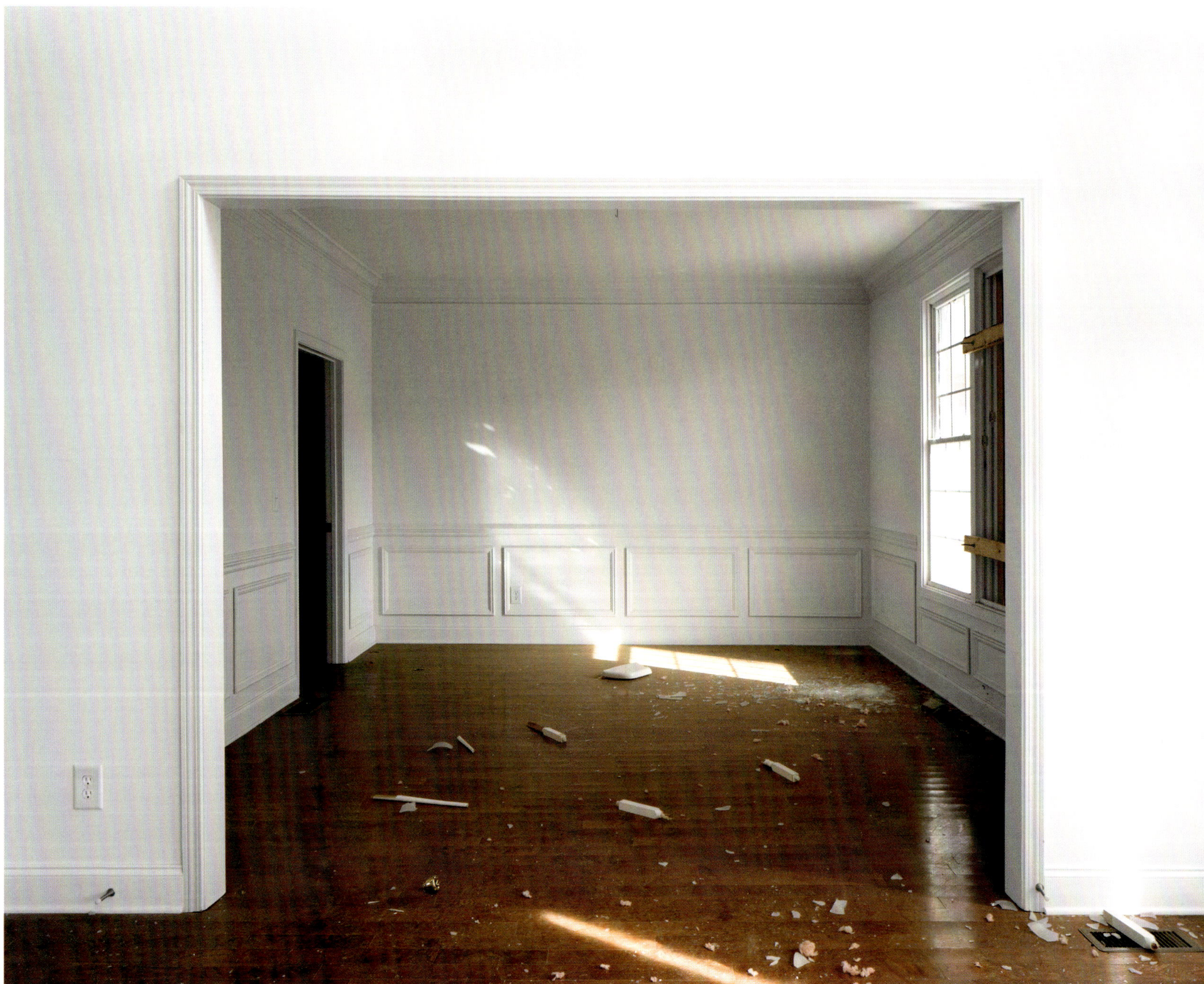

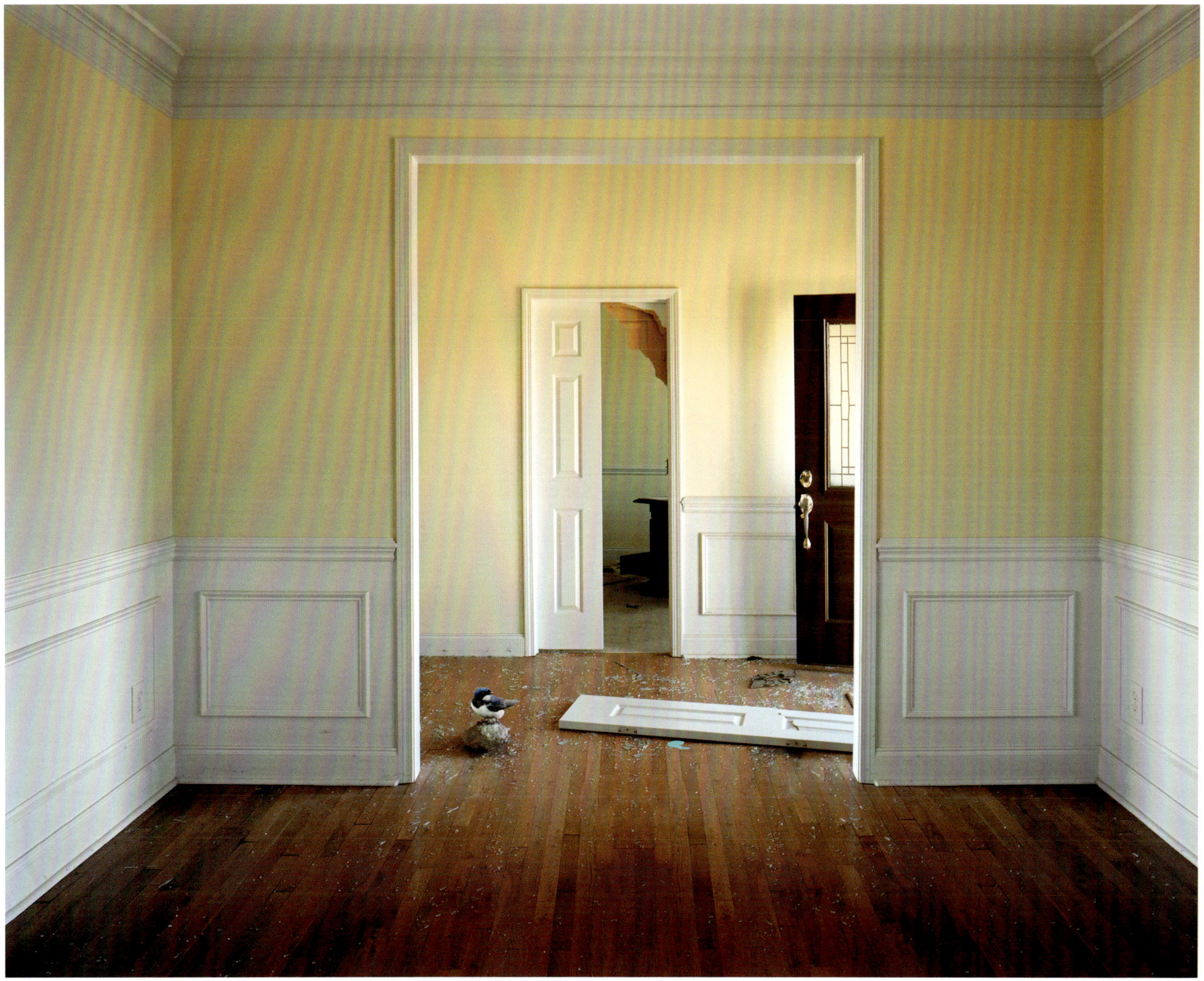

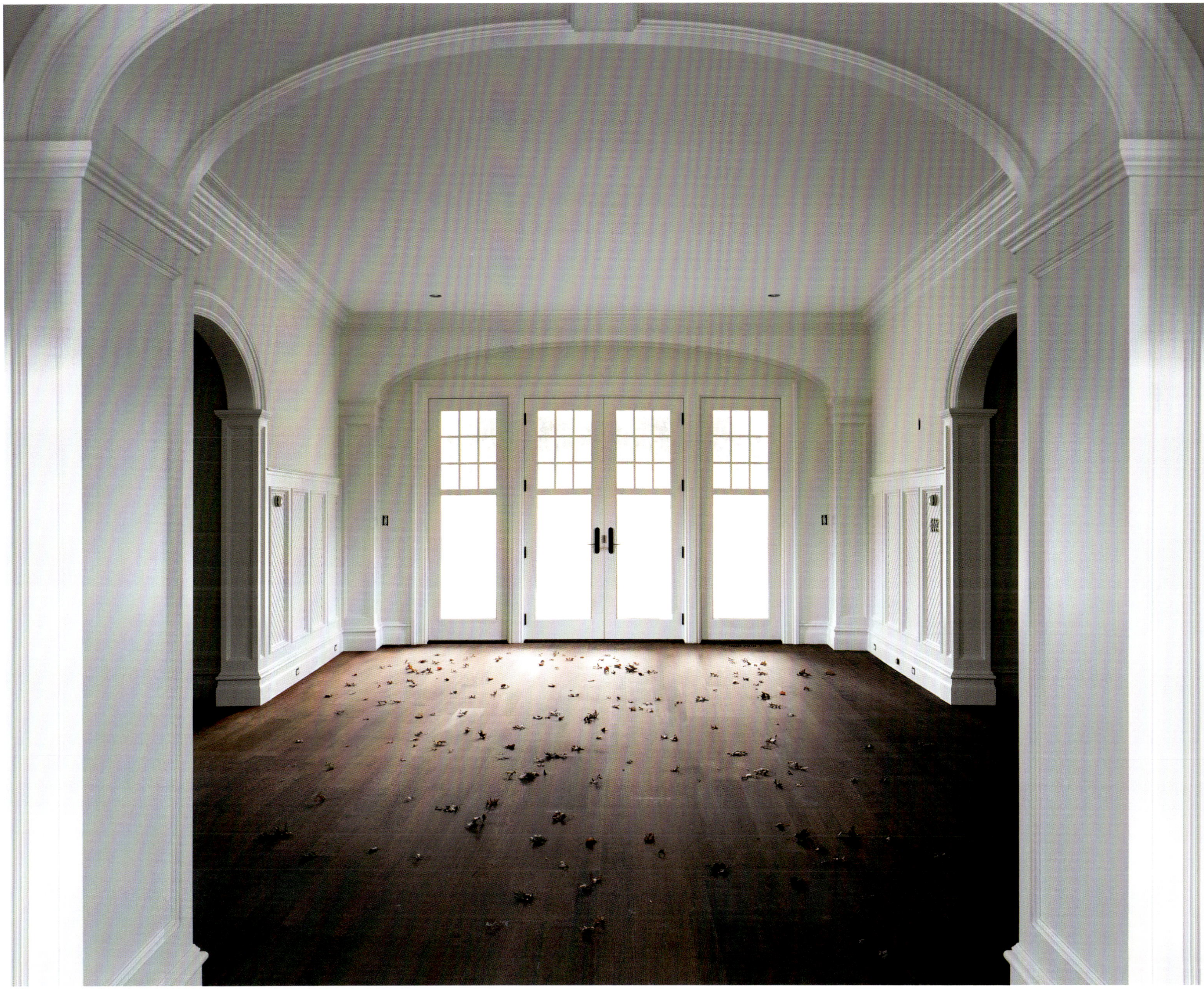

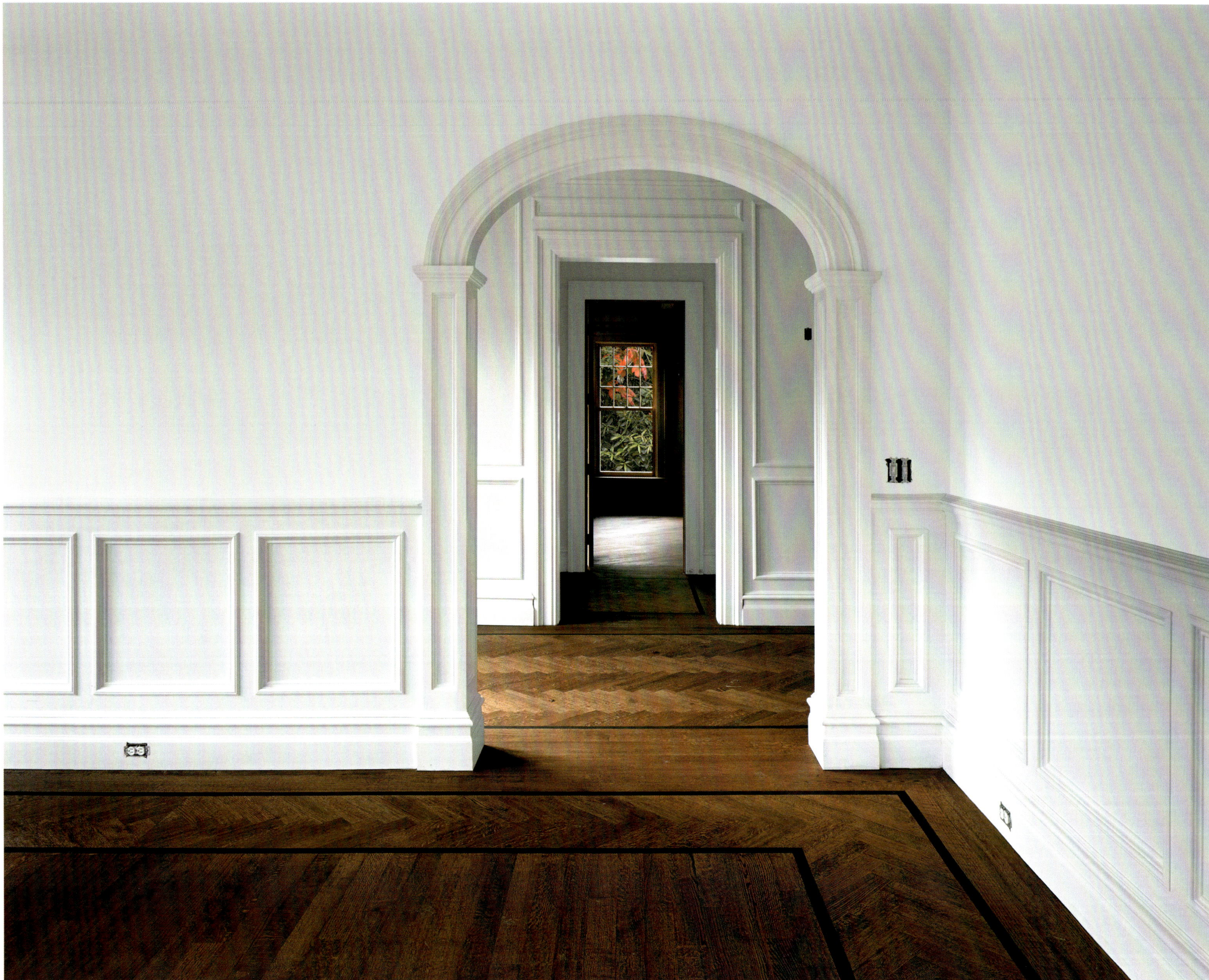

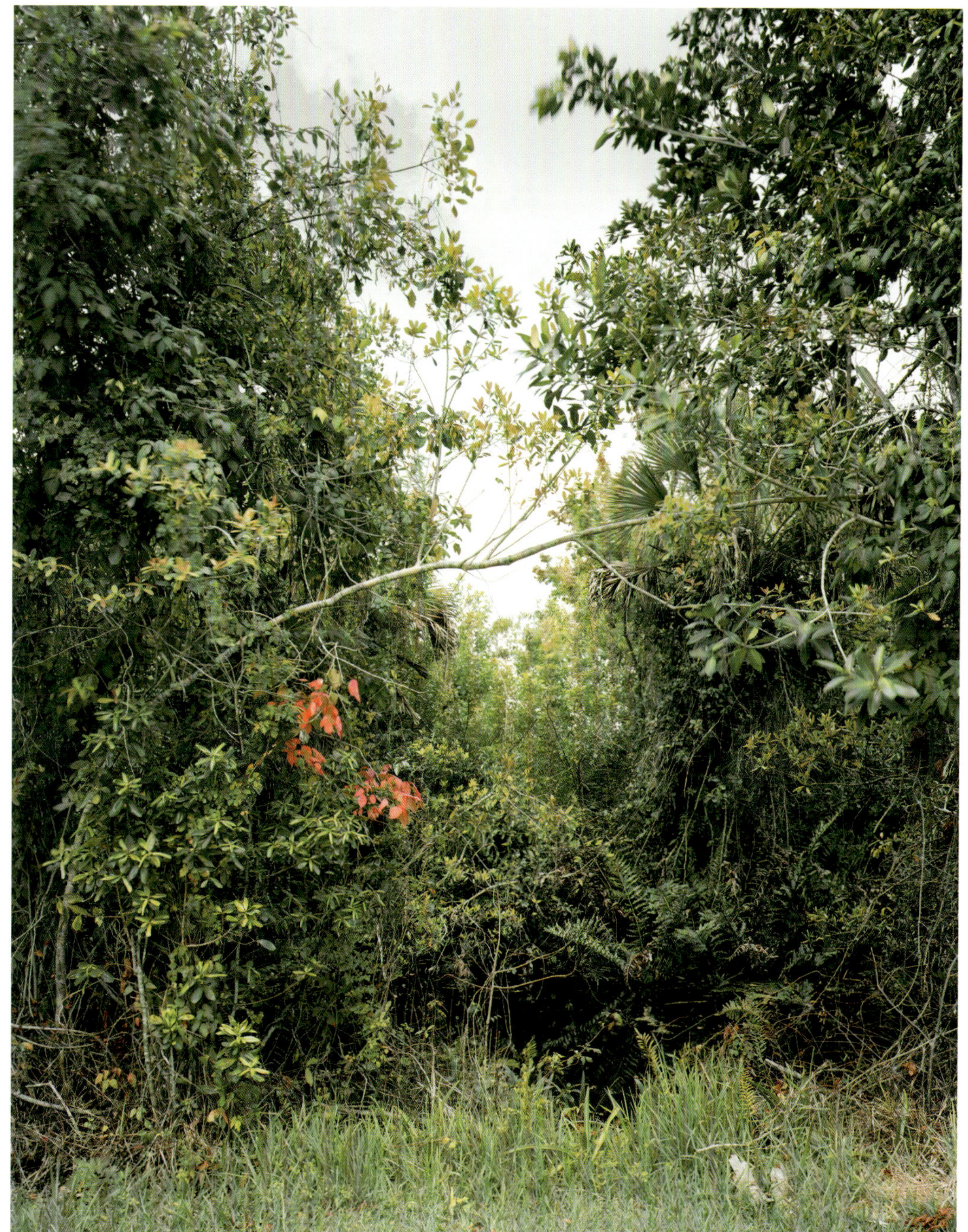

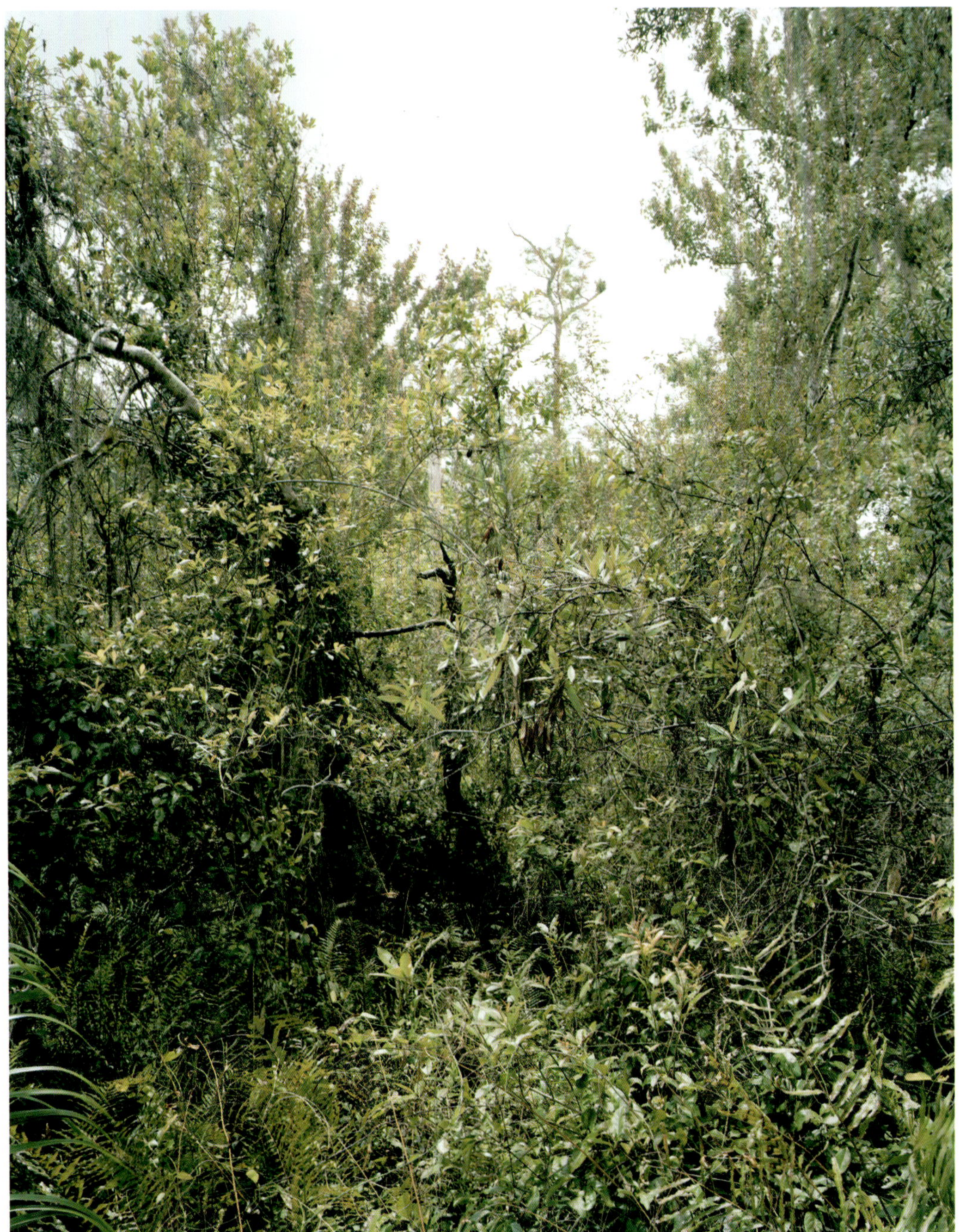

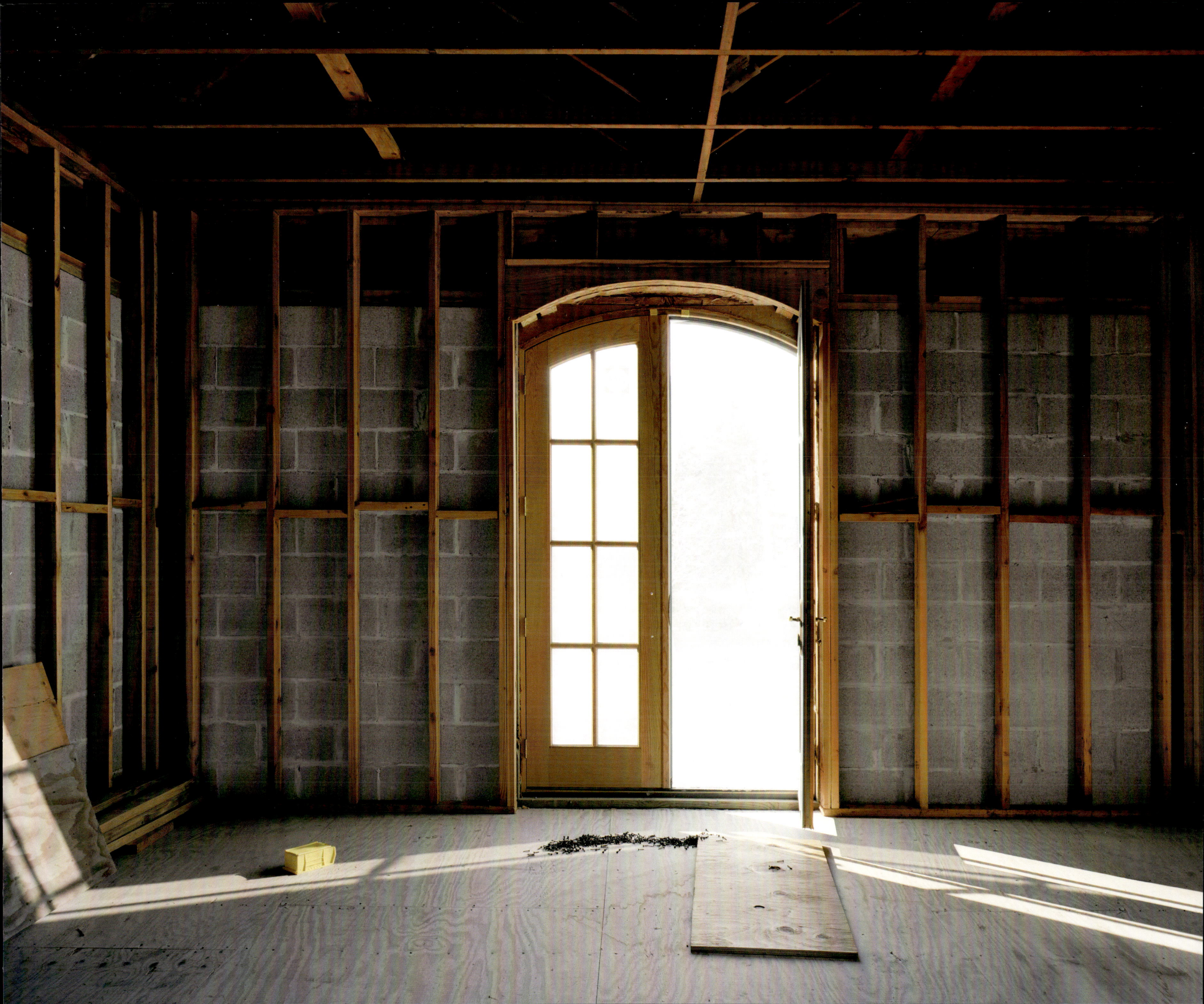

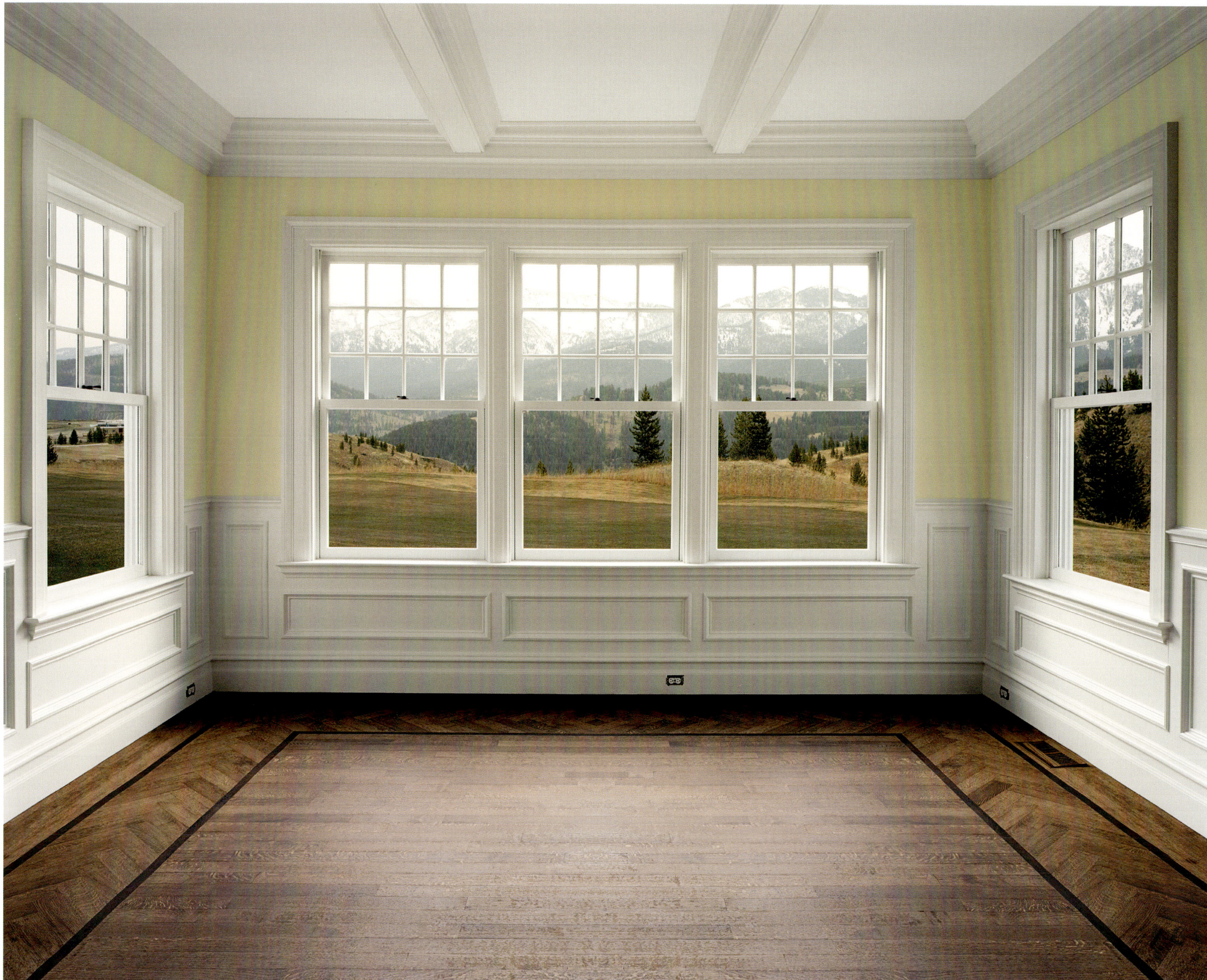

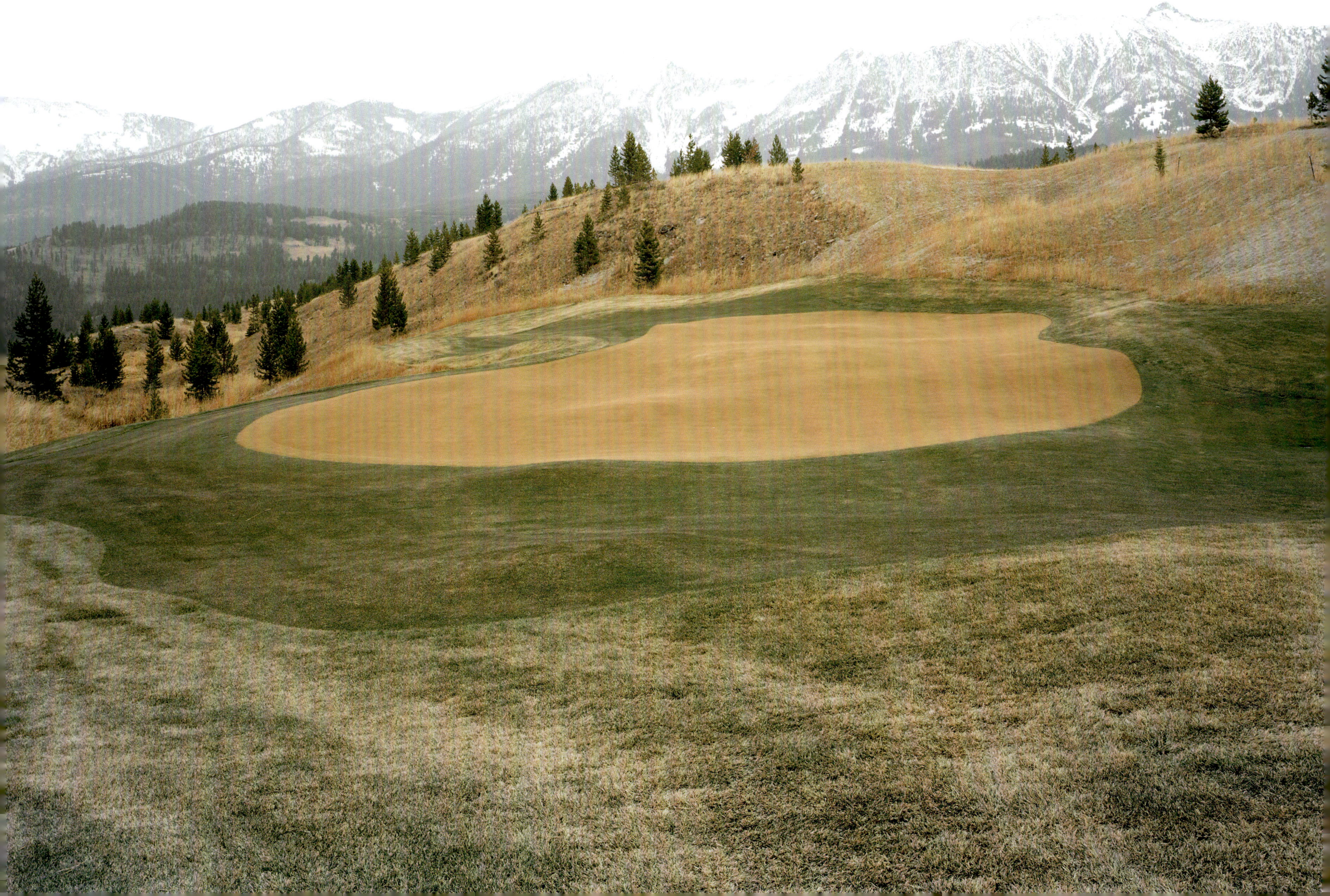

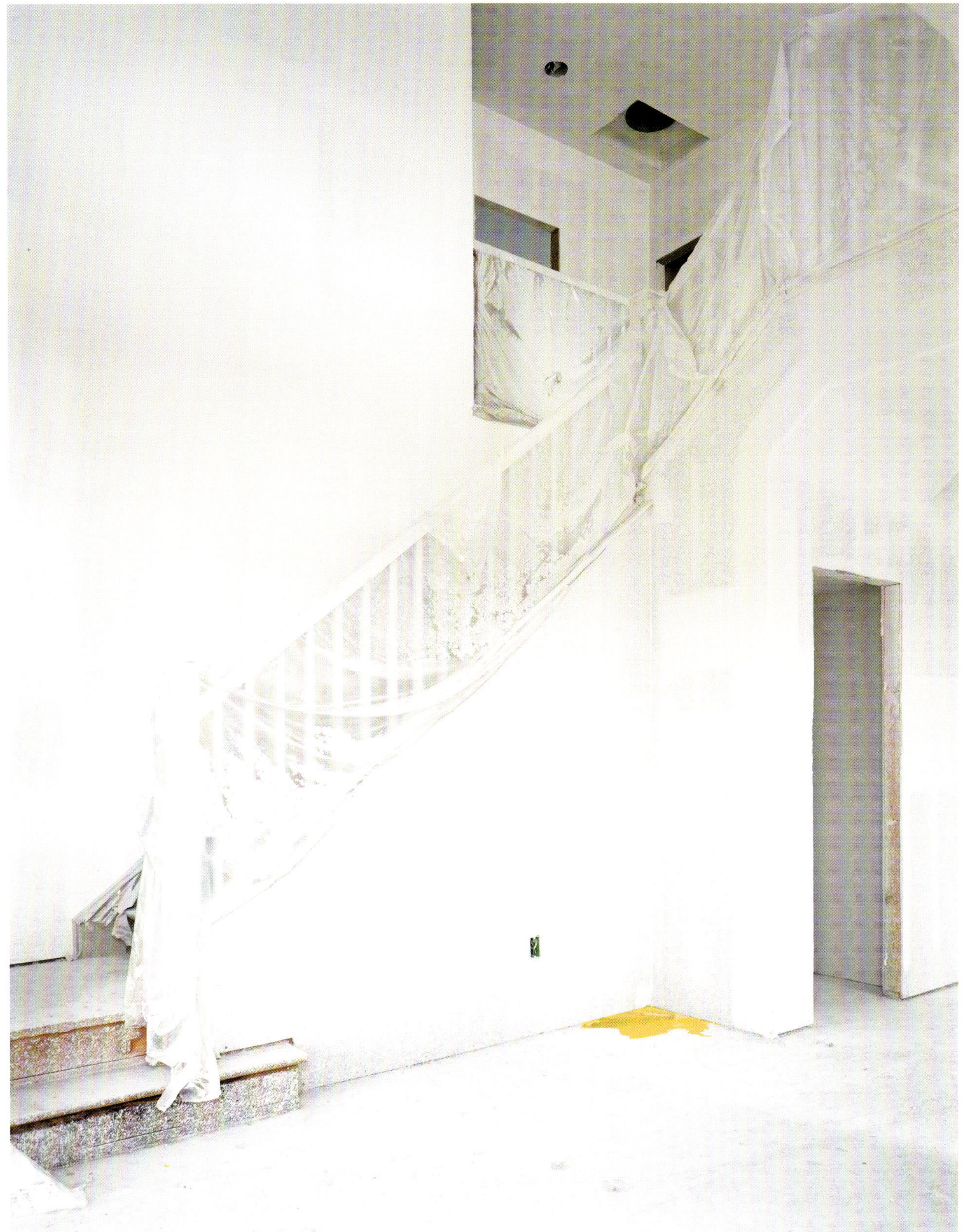

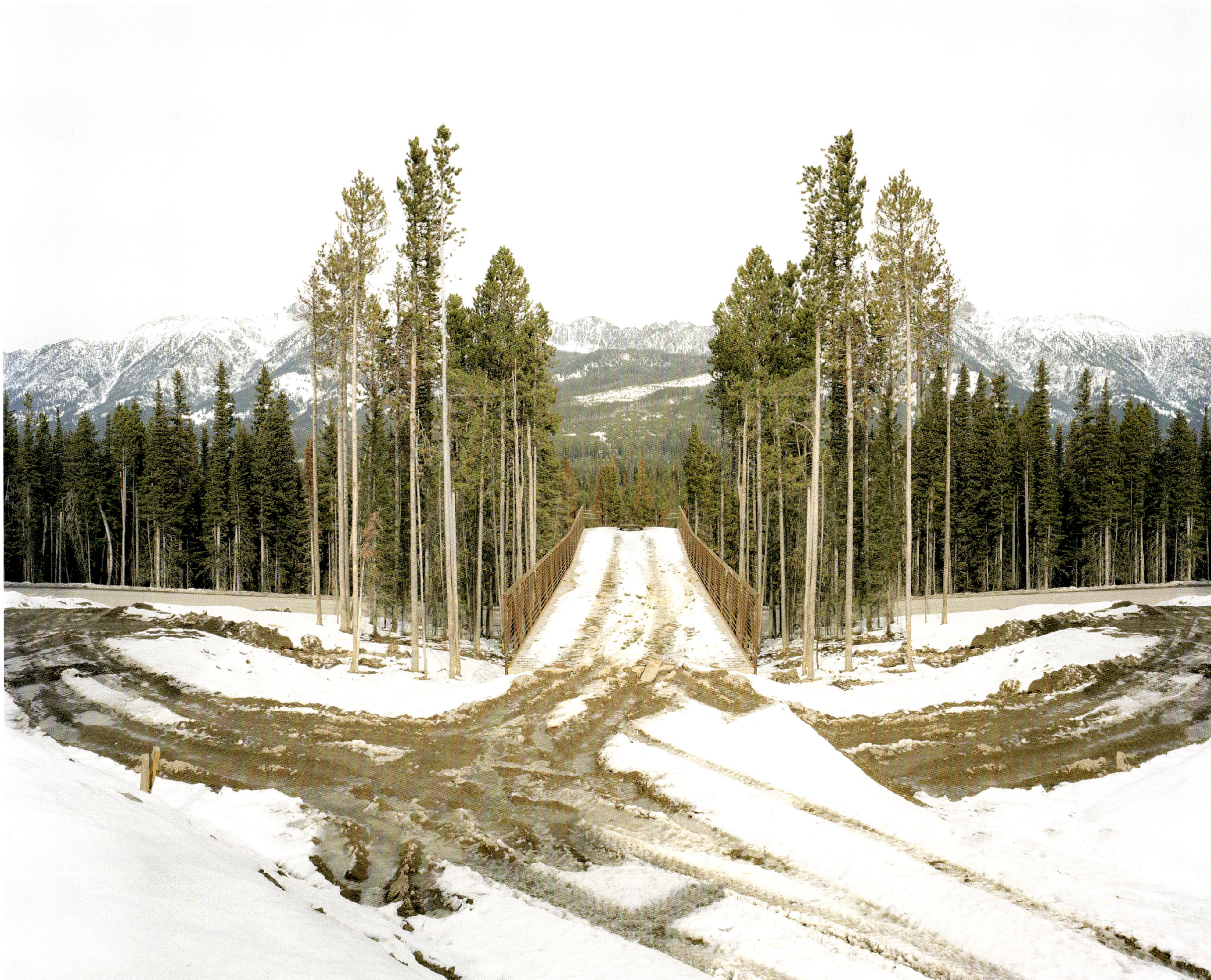

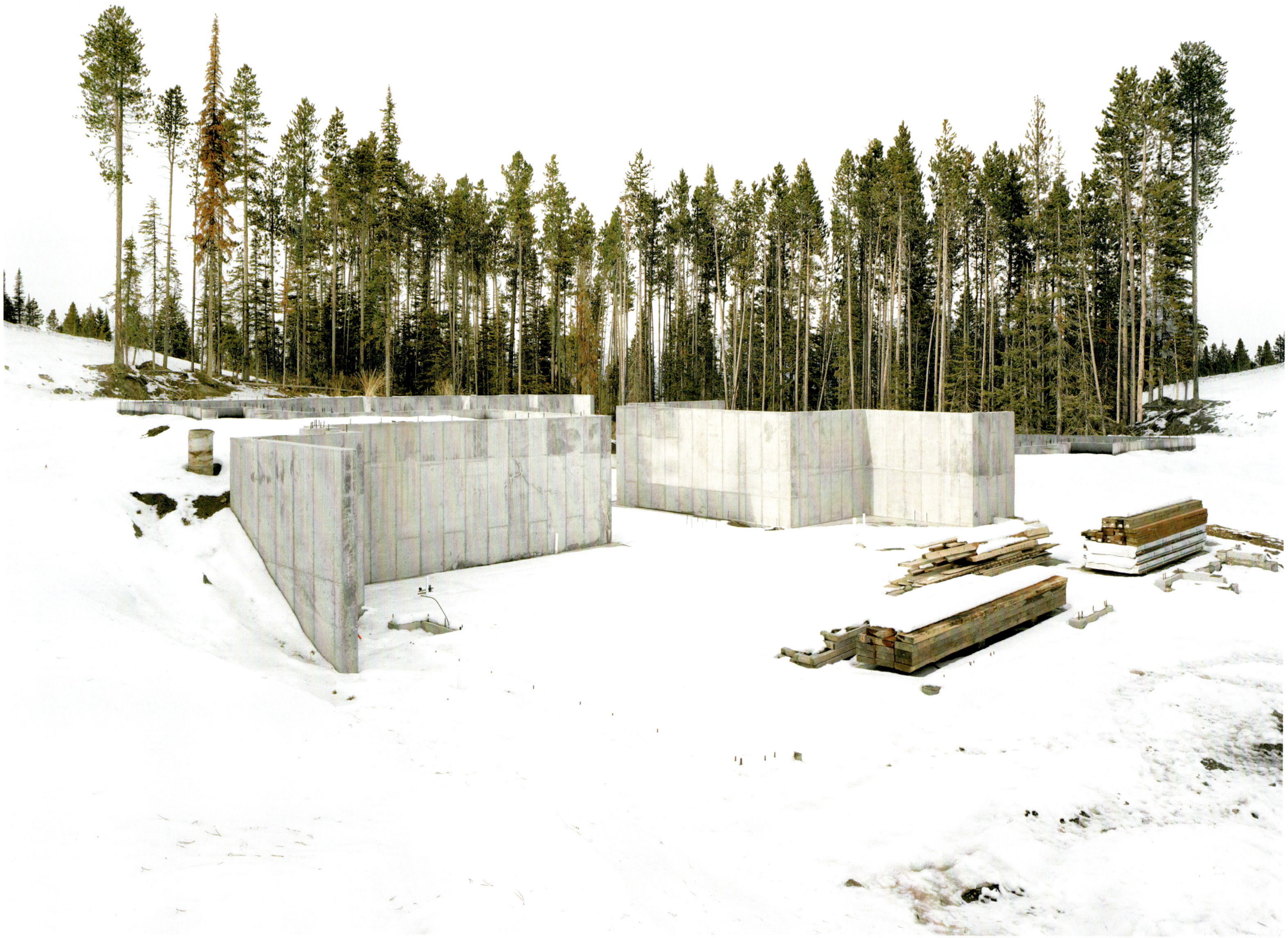

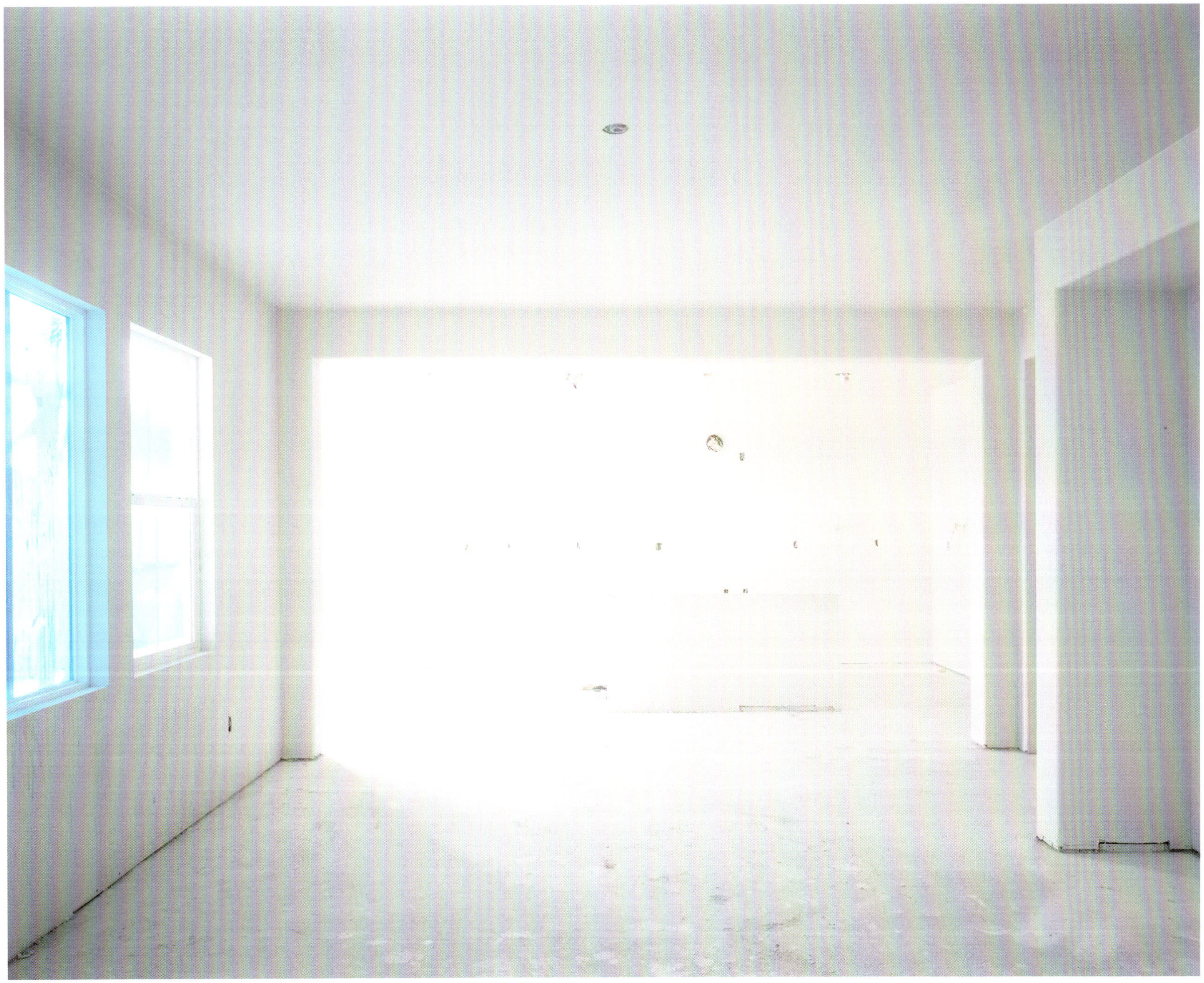

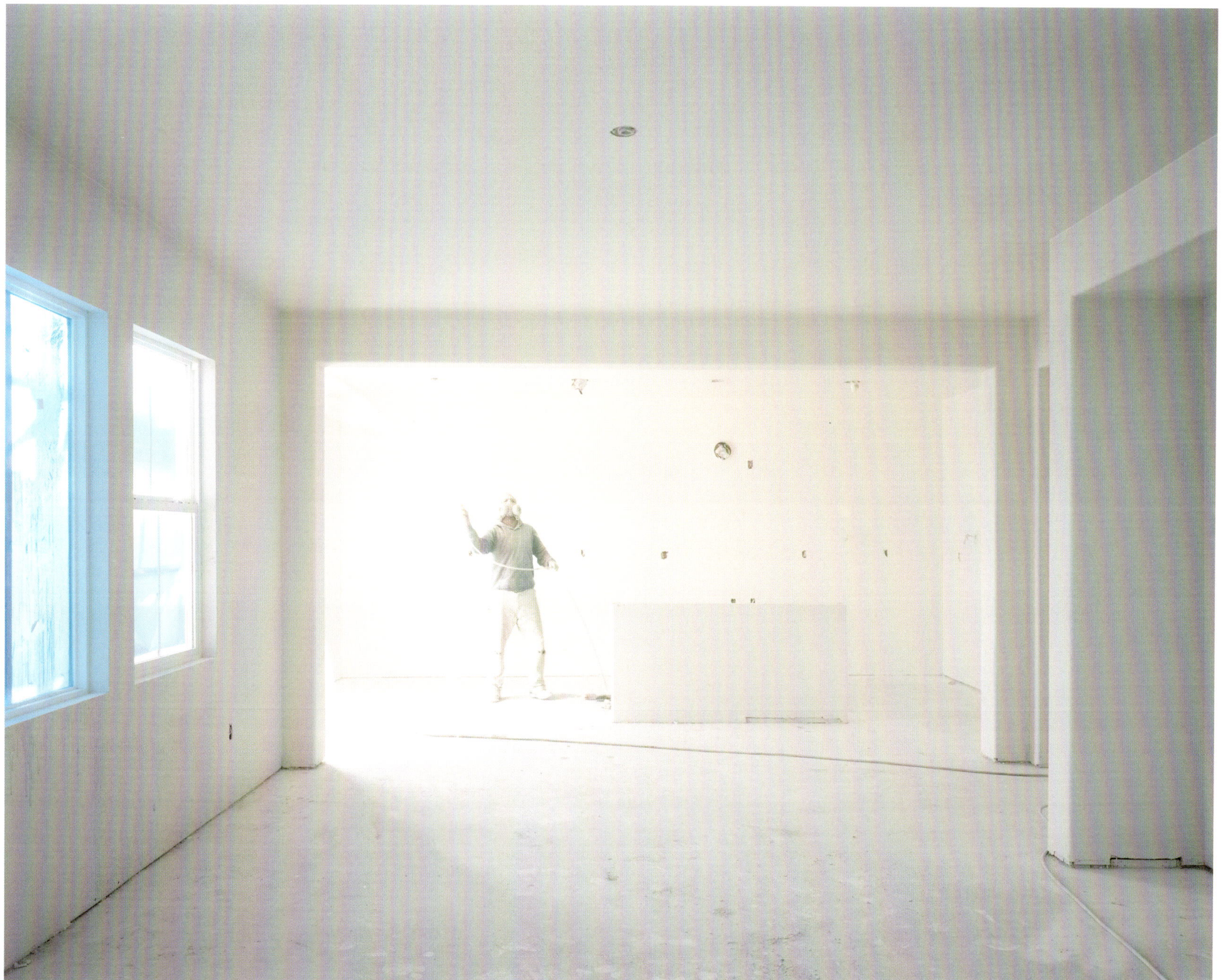

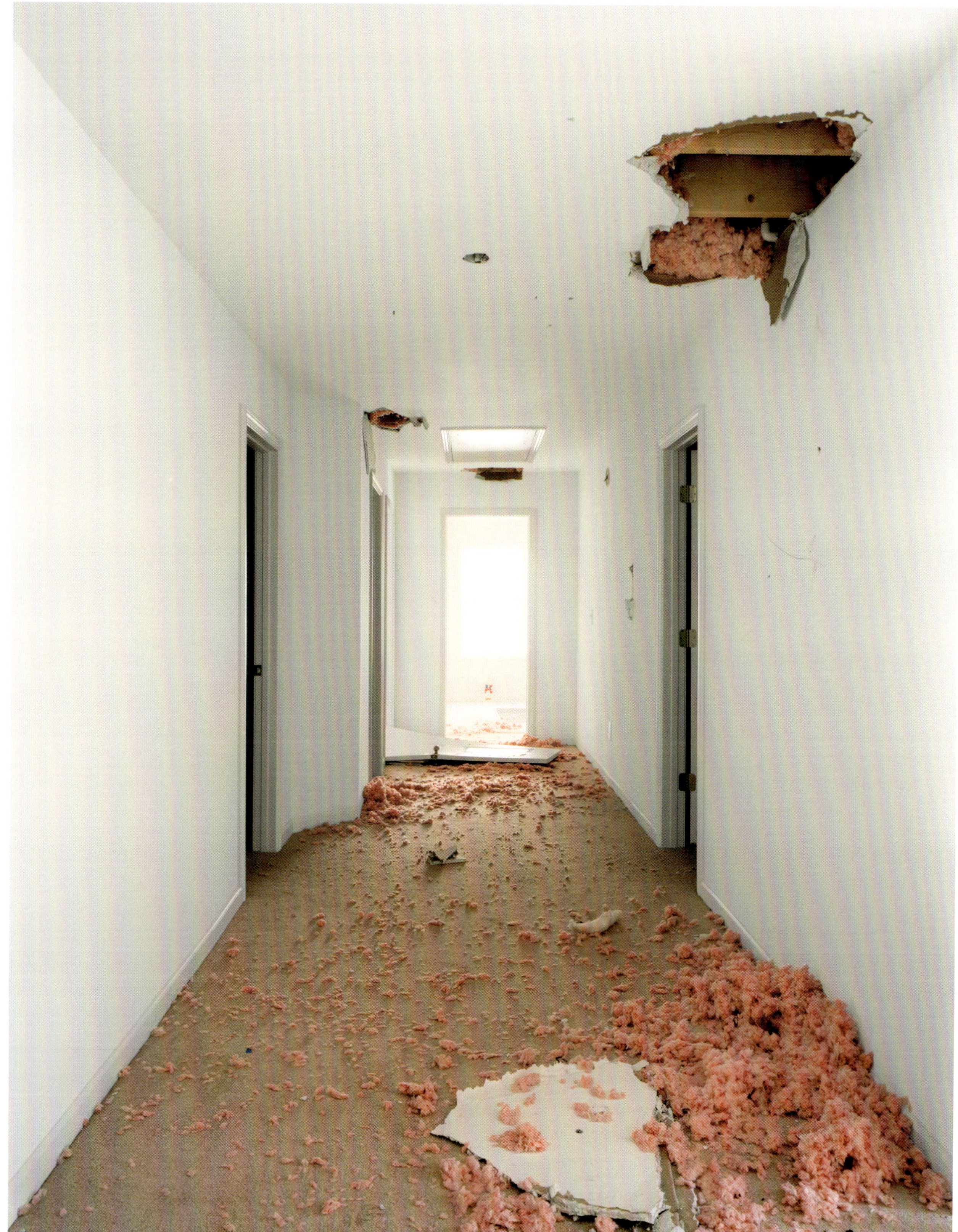

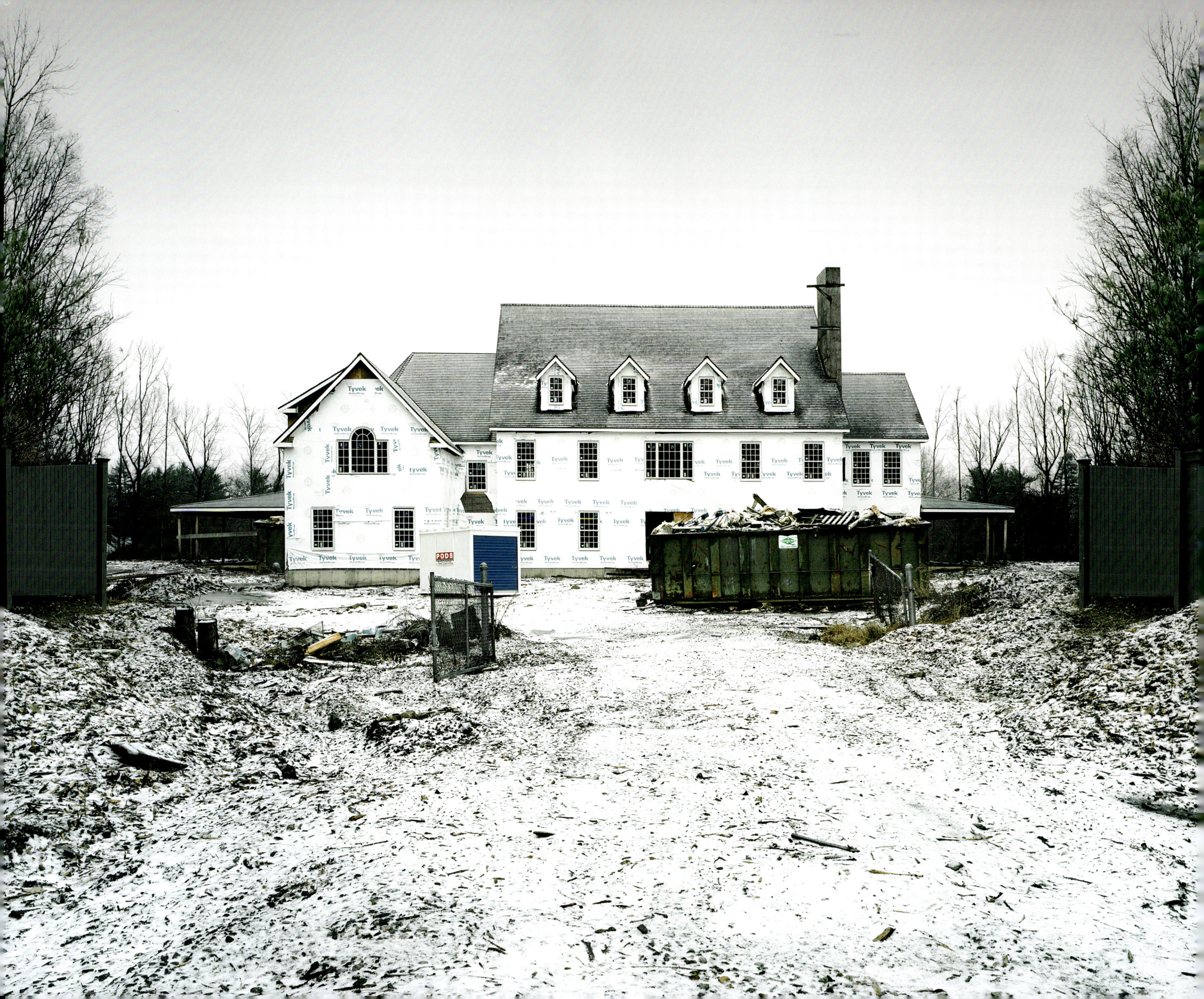
Tyvek
PODS

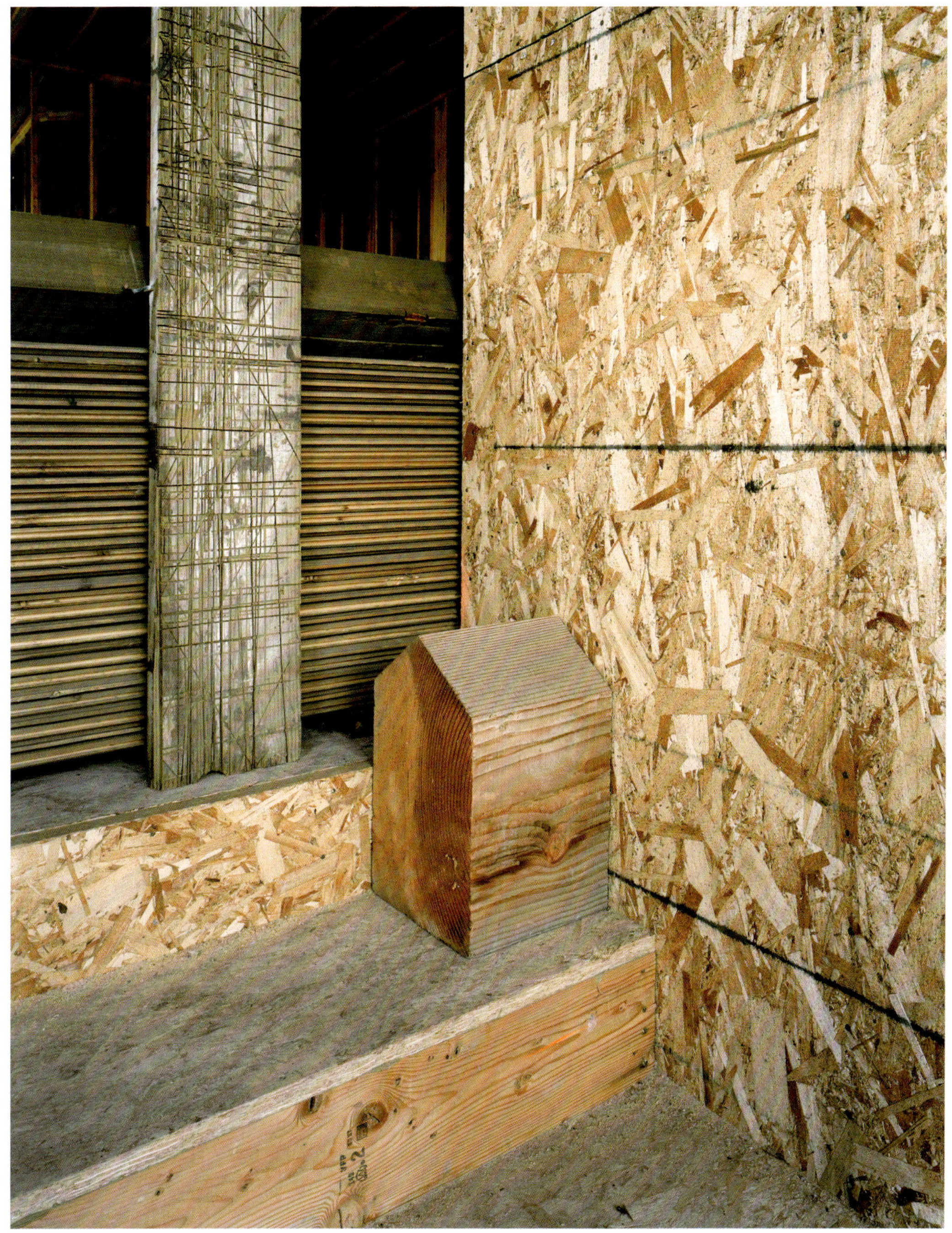

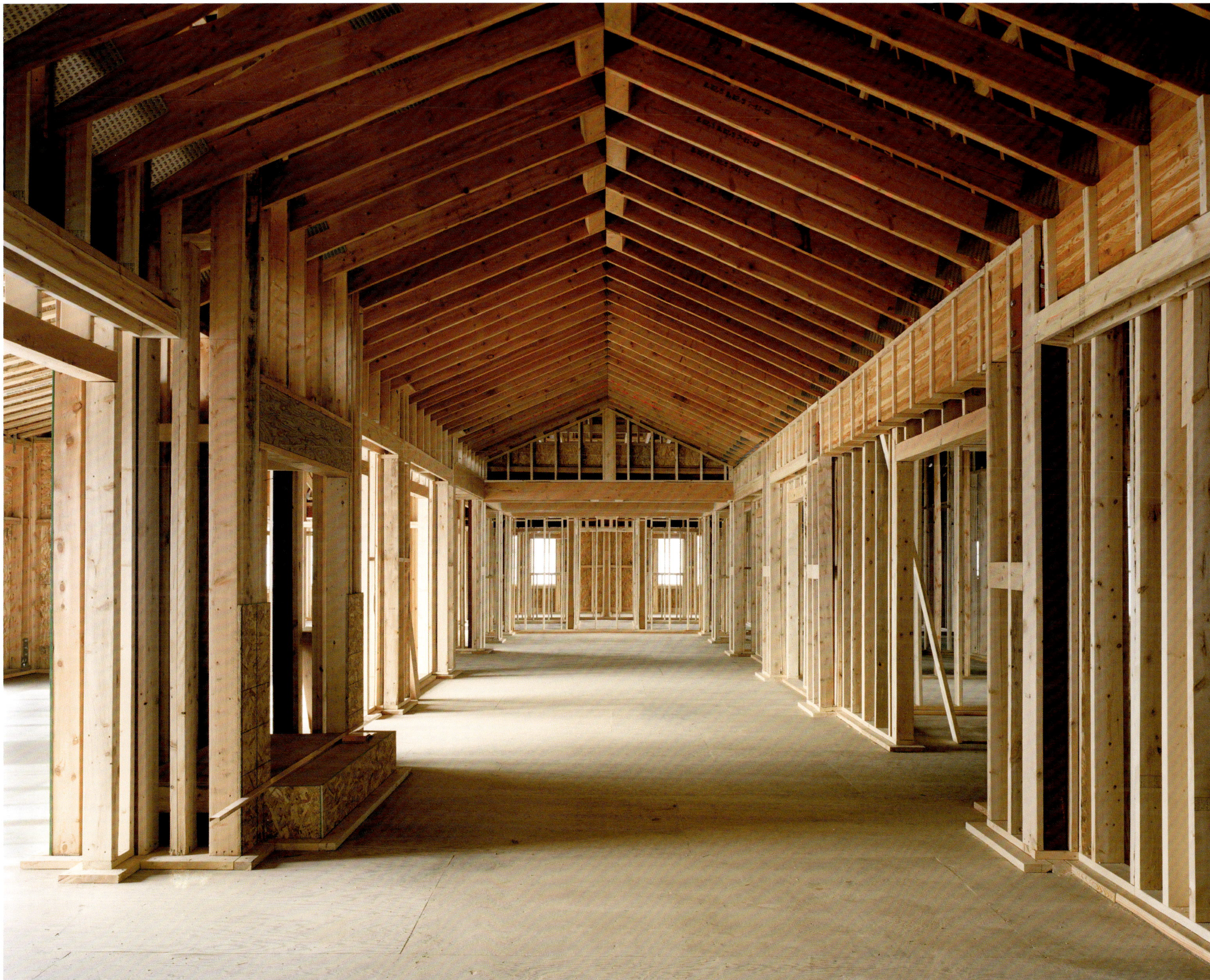

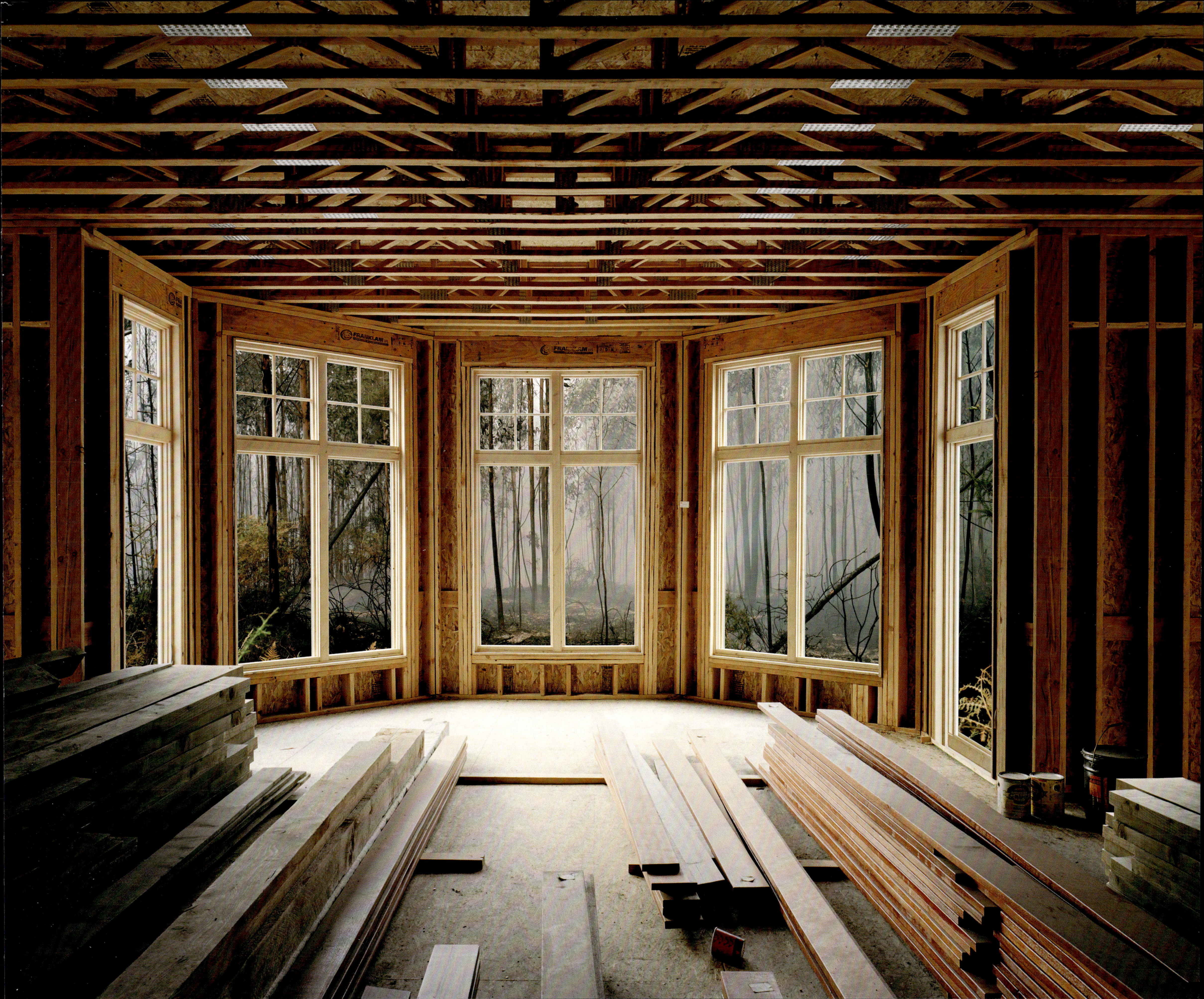

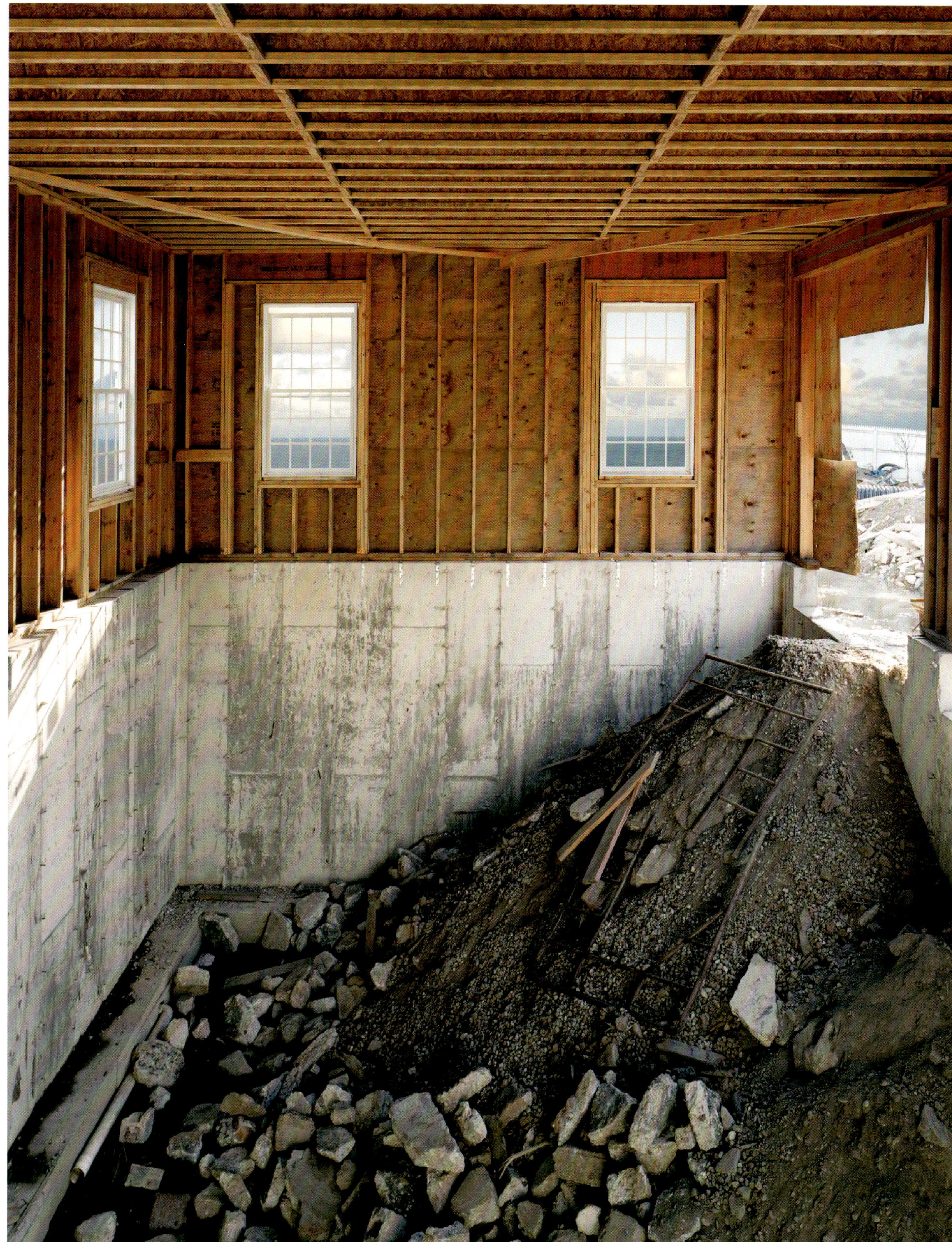

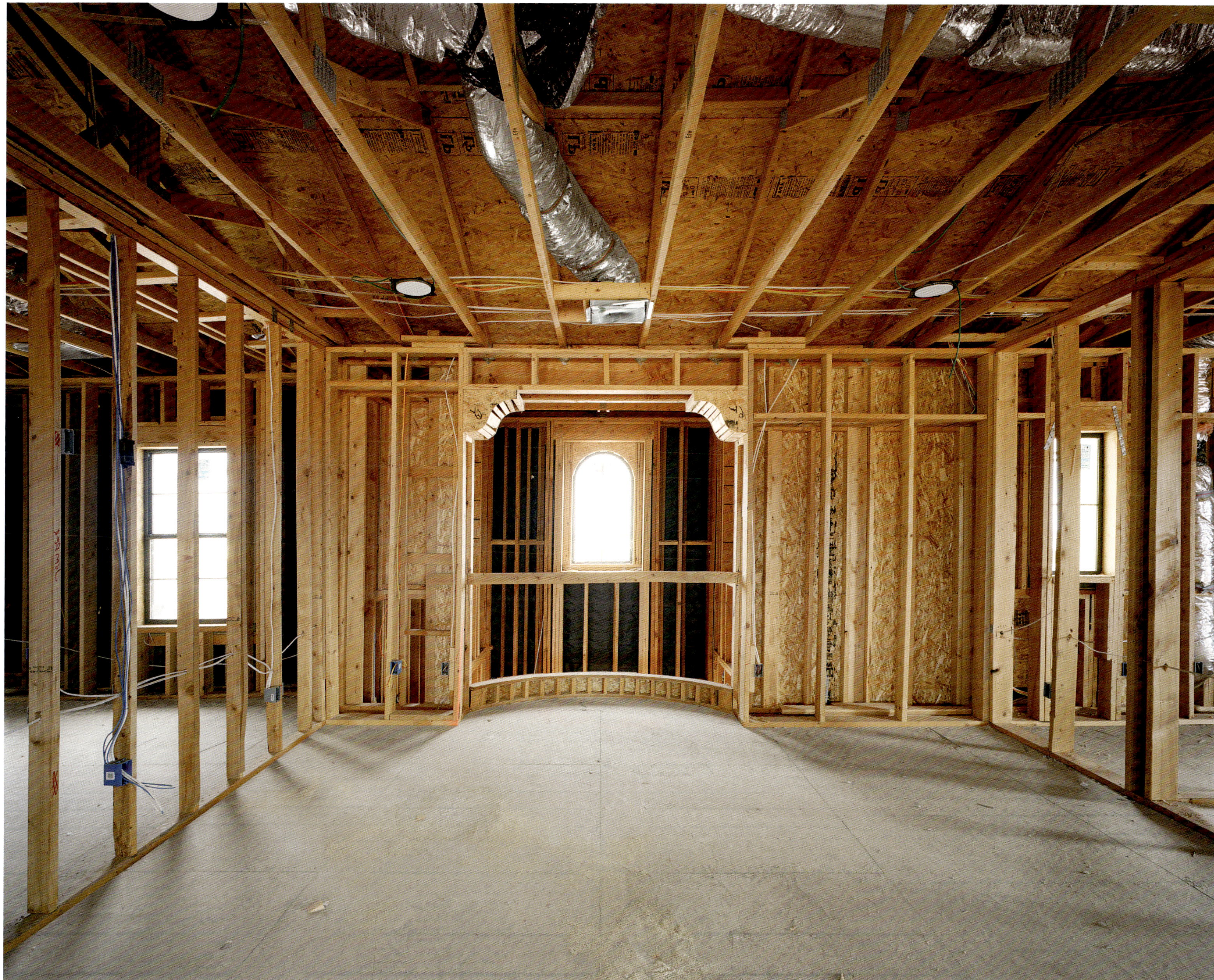

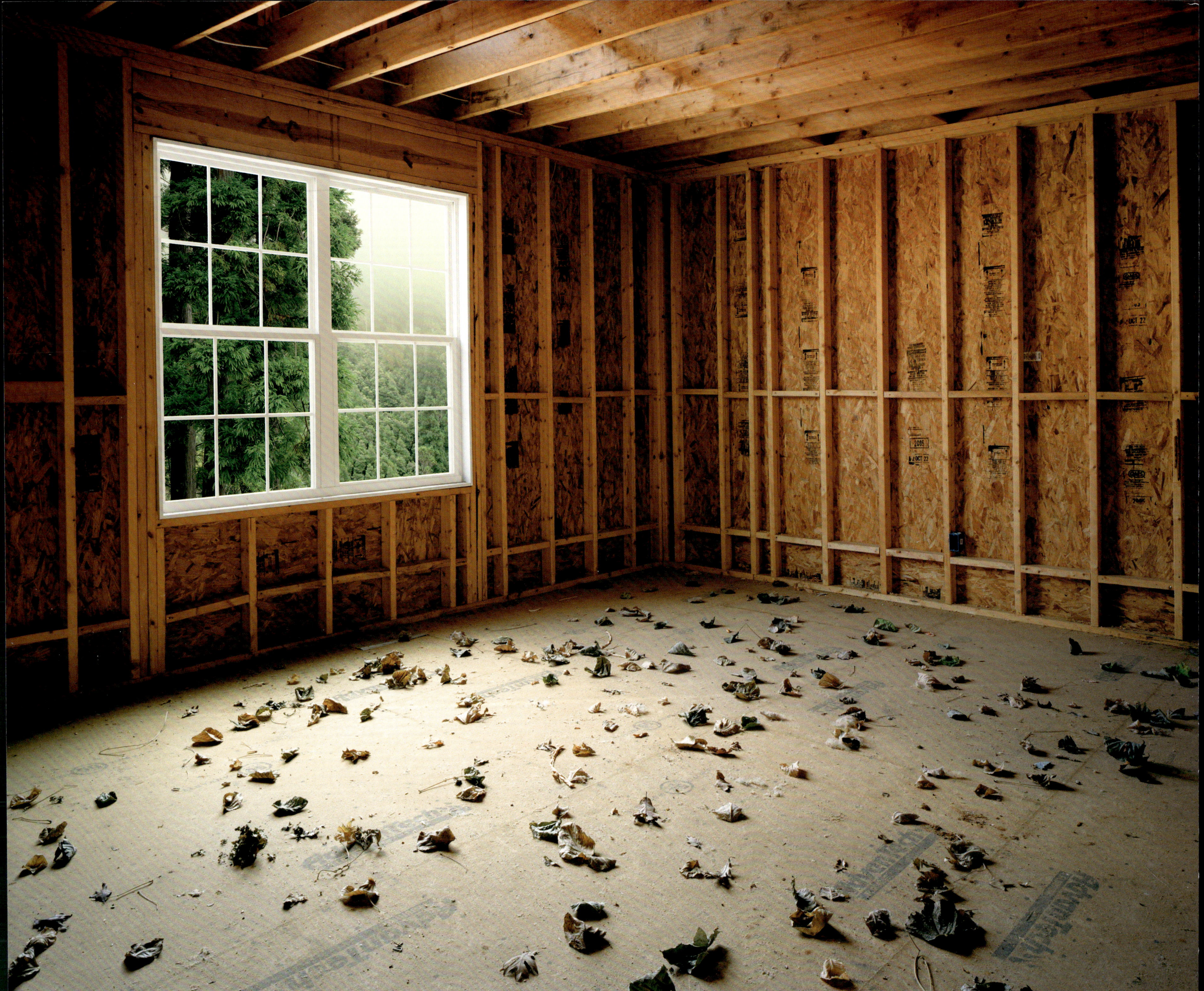

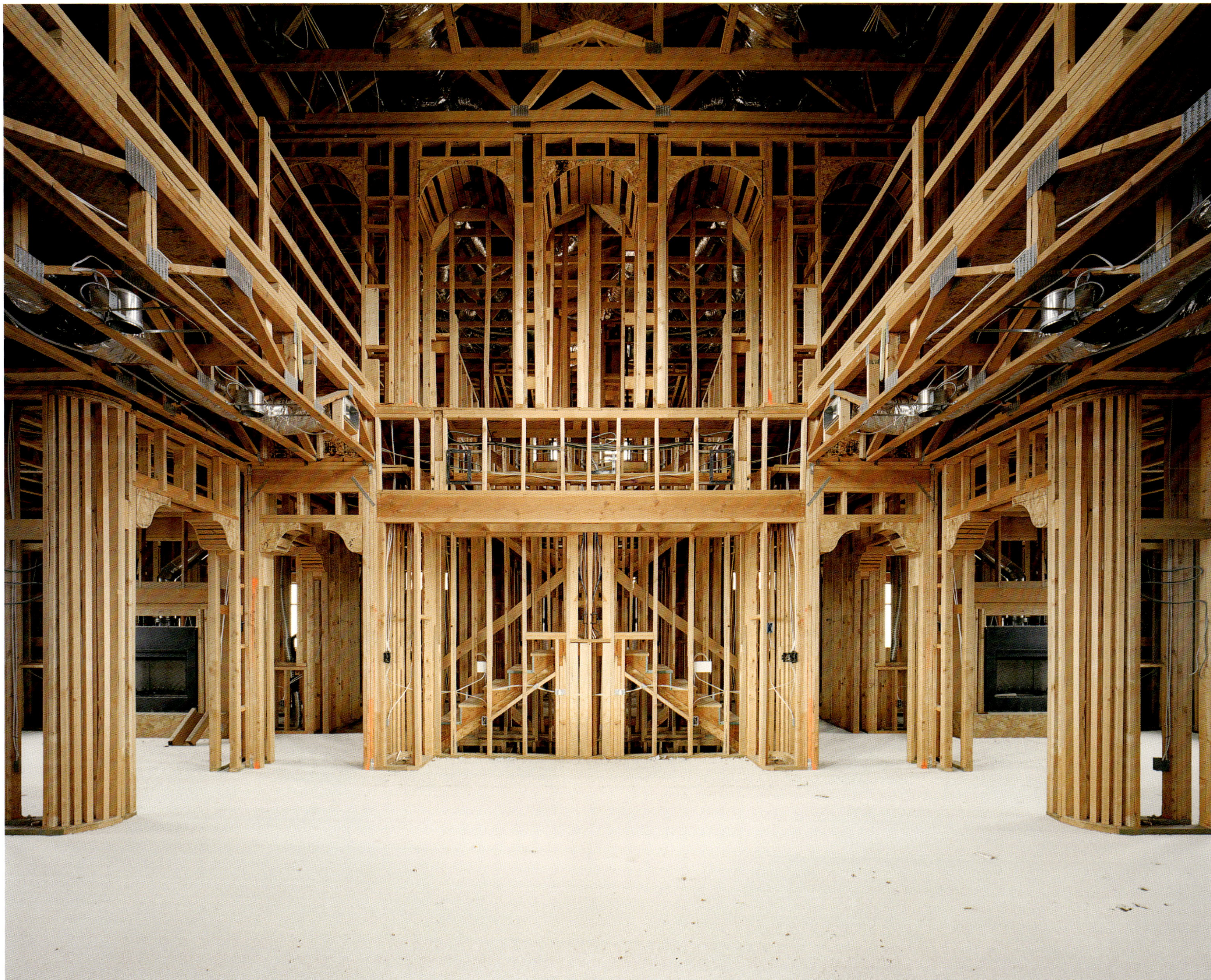

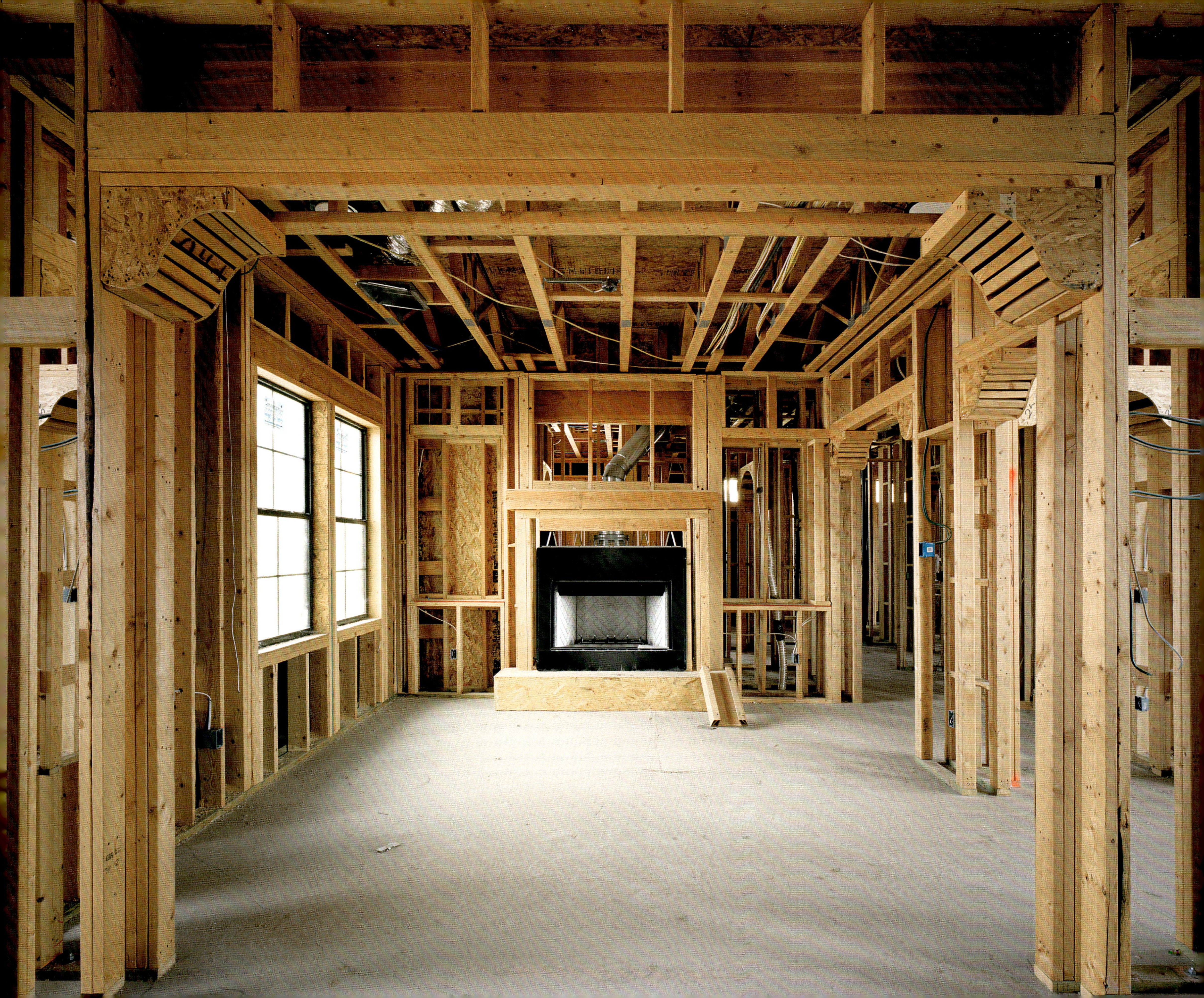

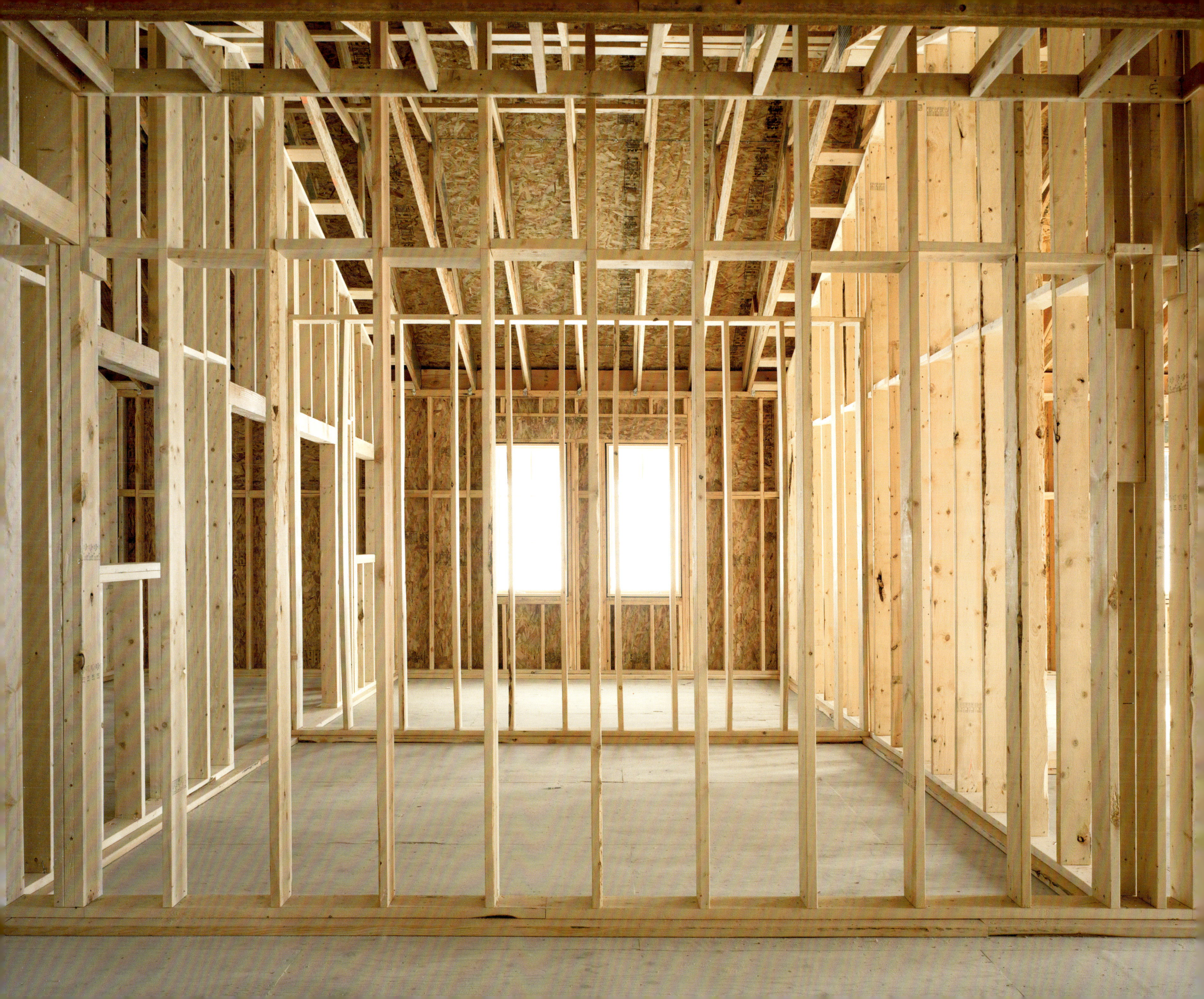

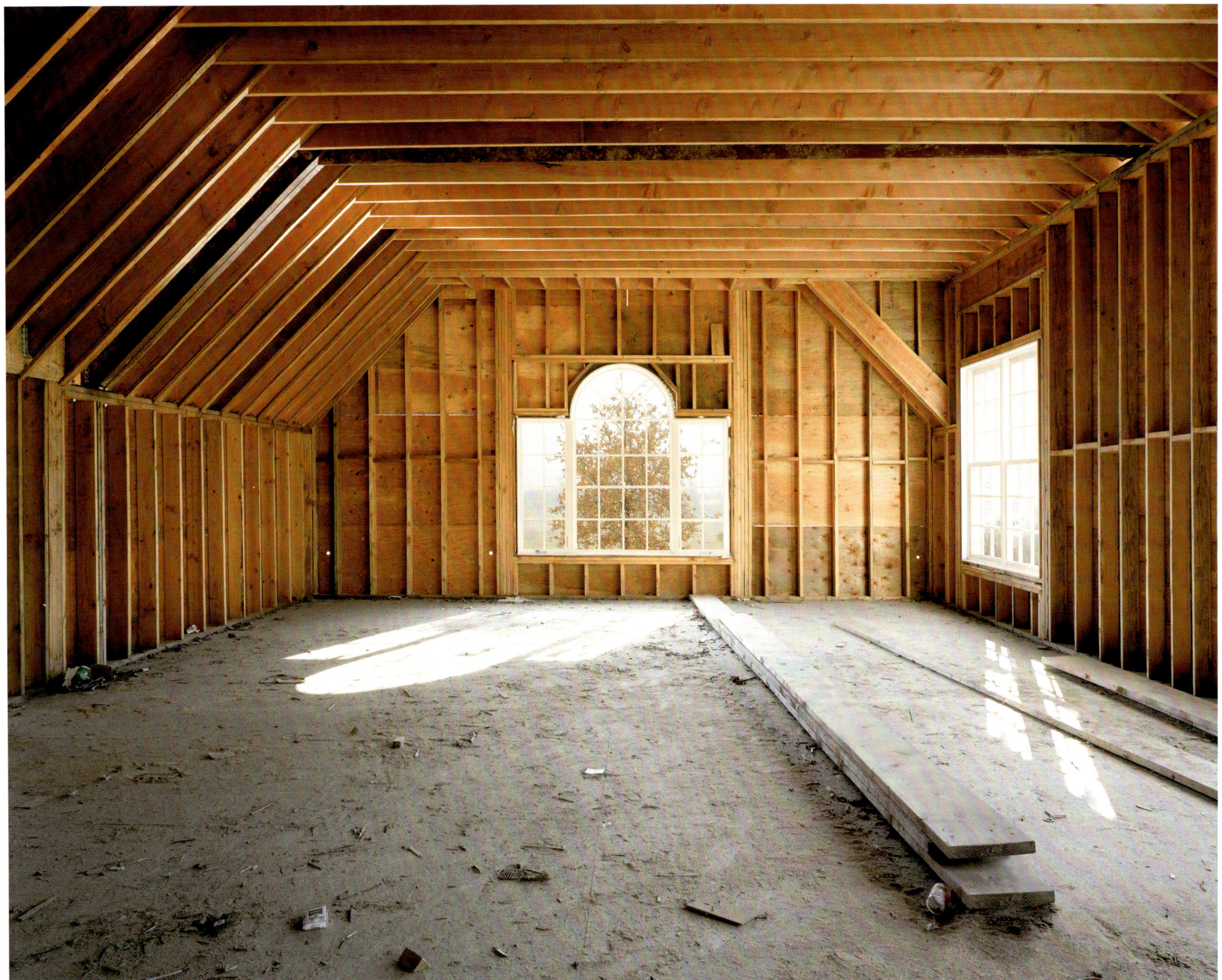

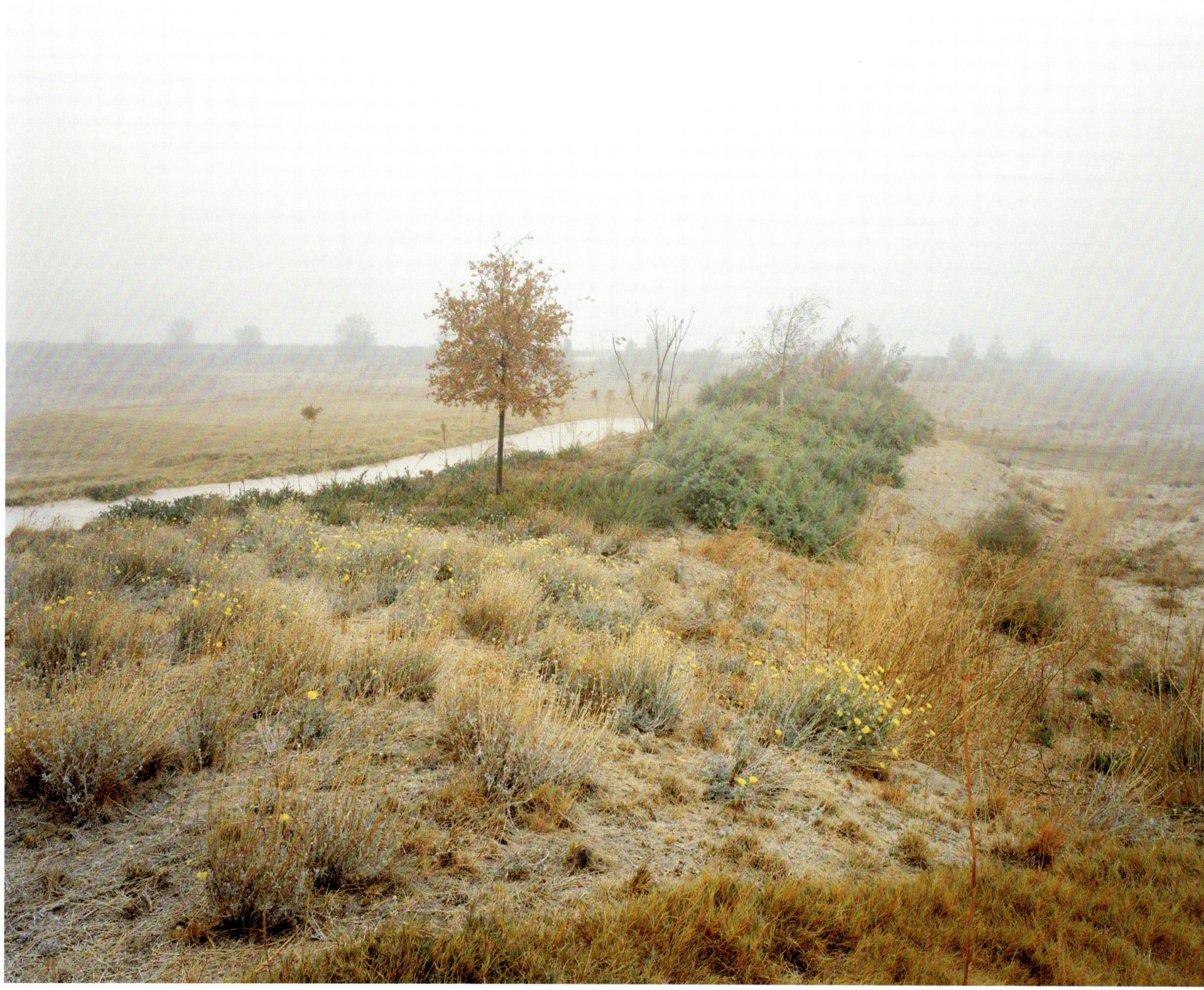

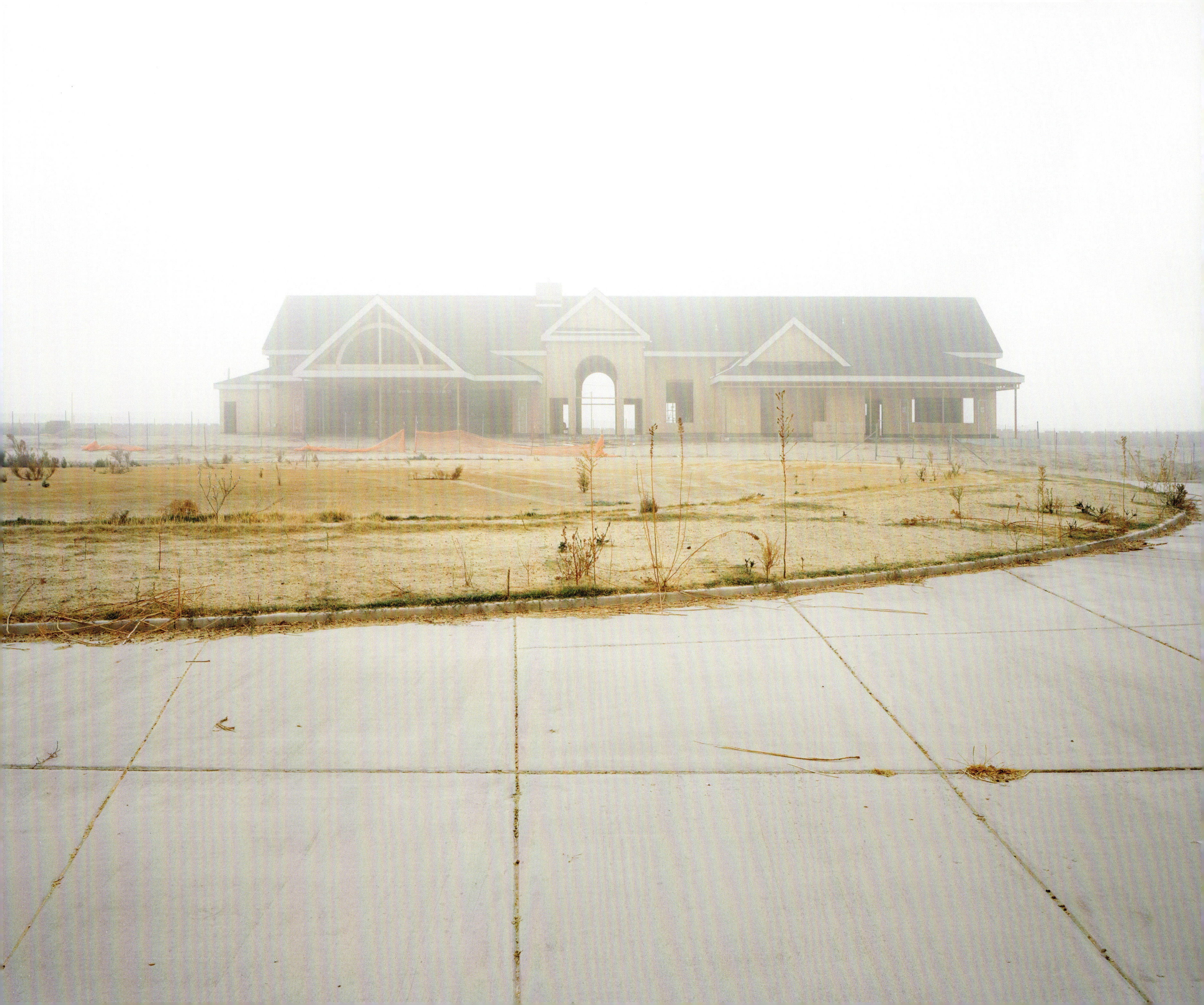

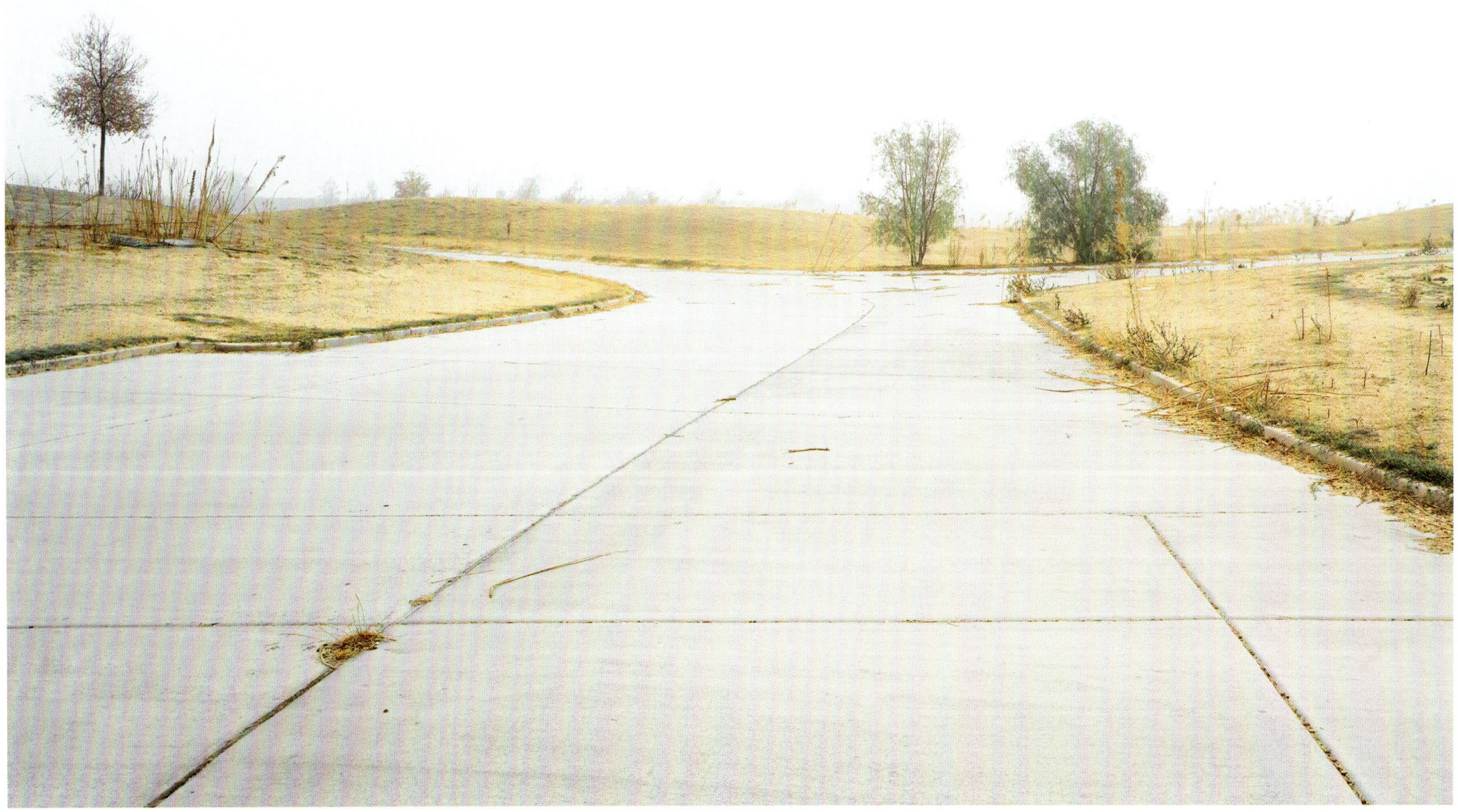

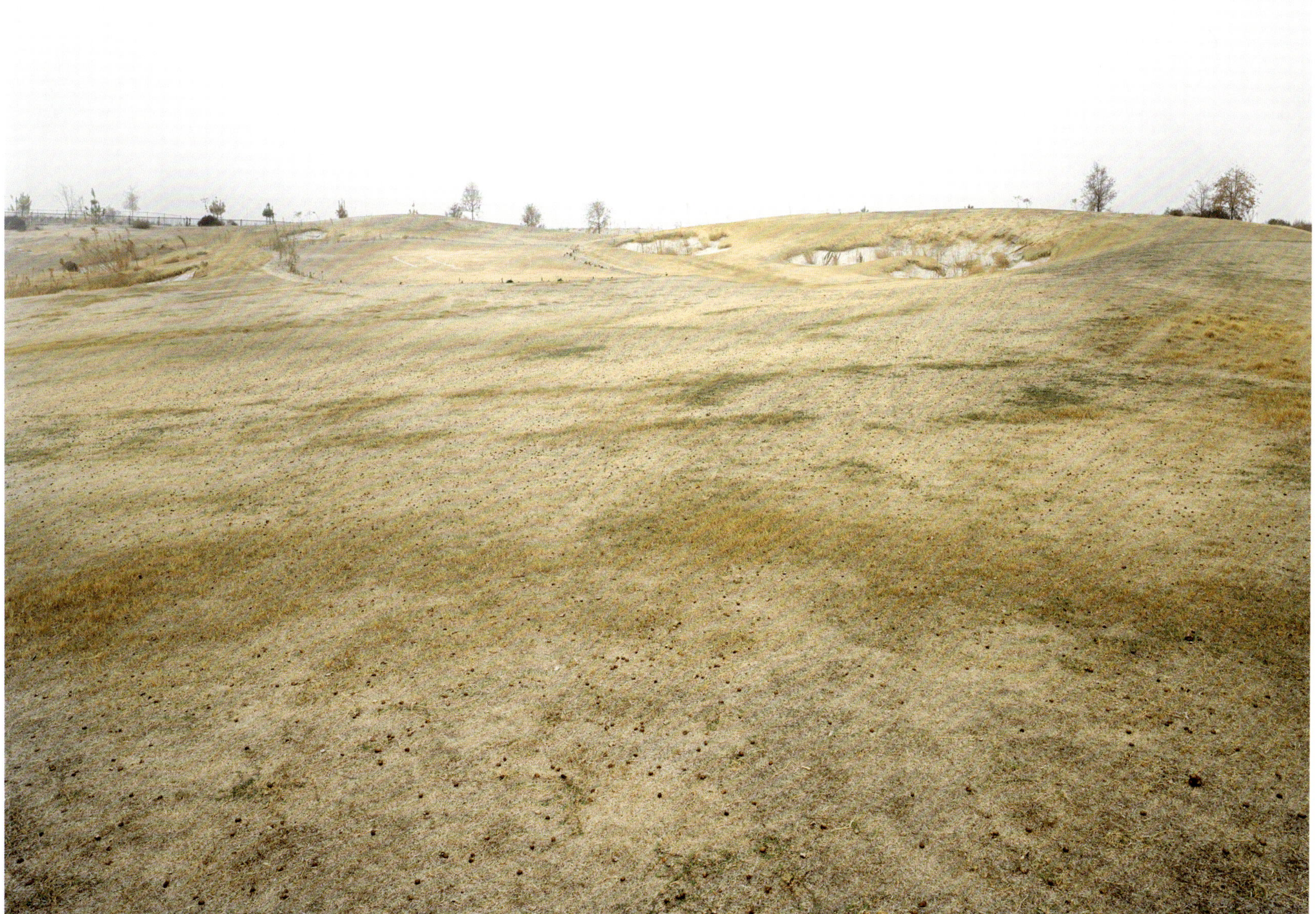

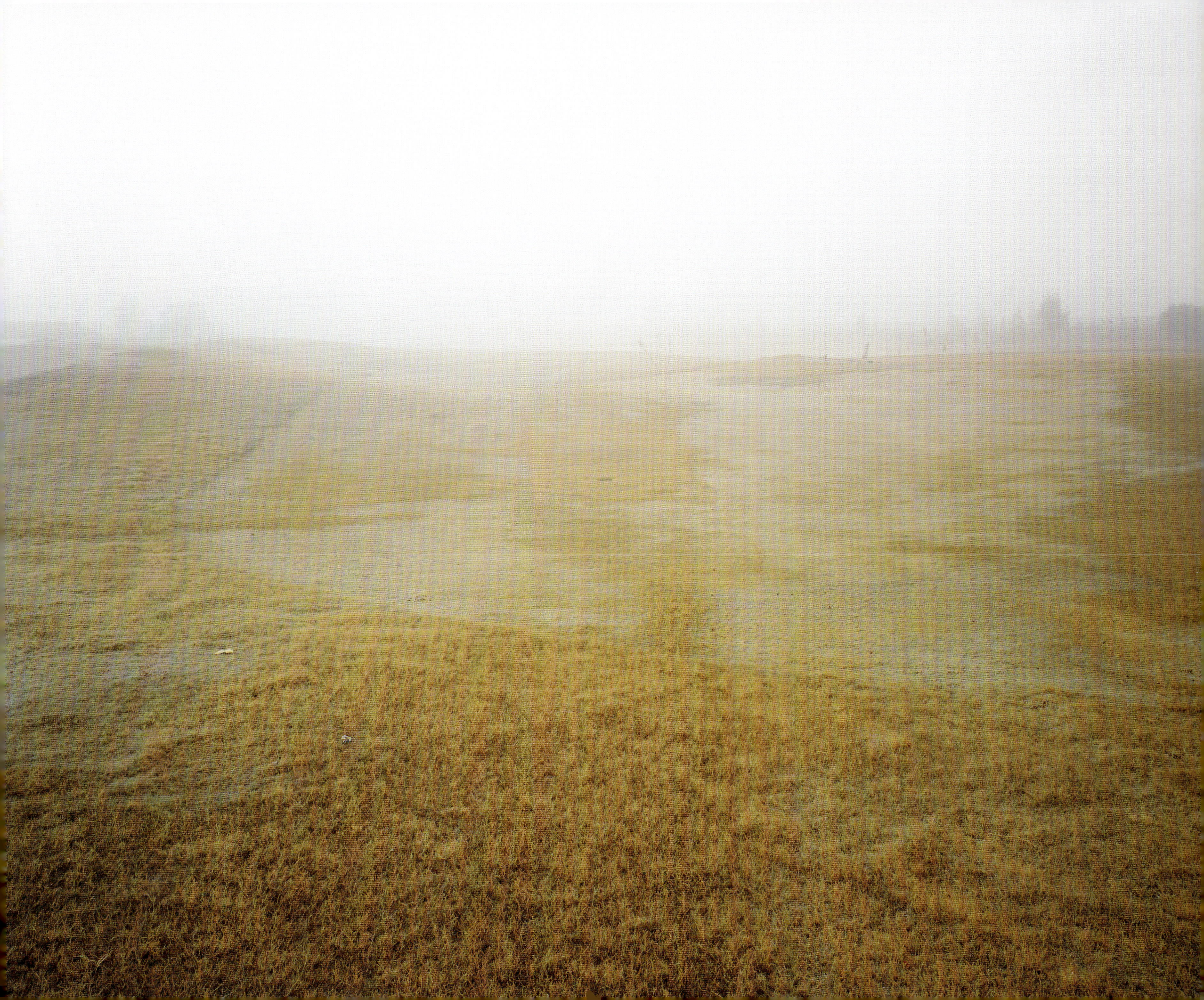

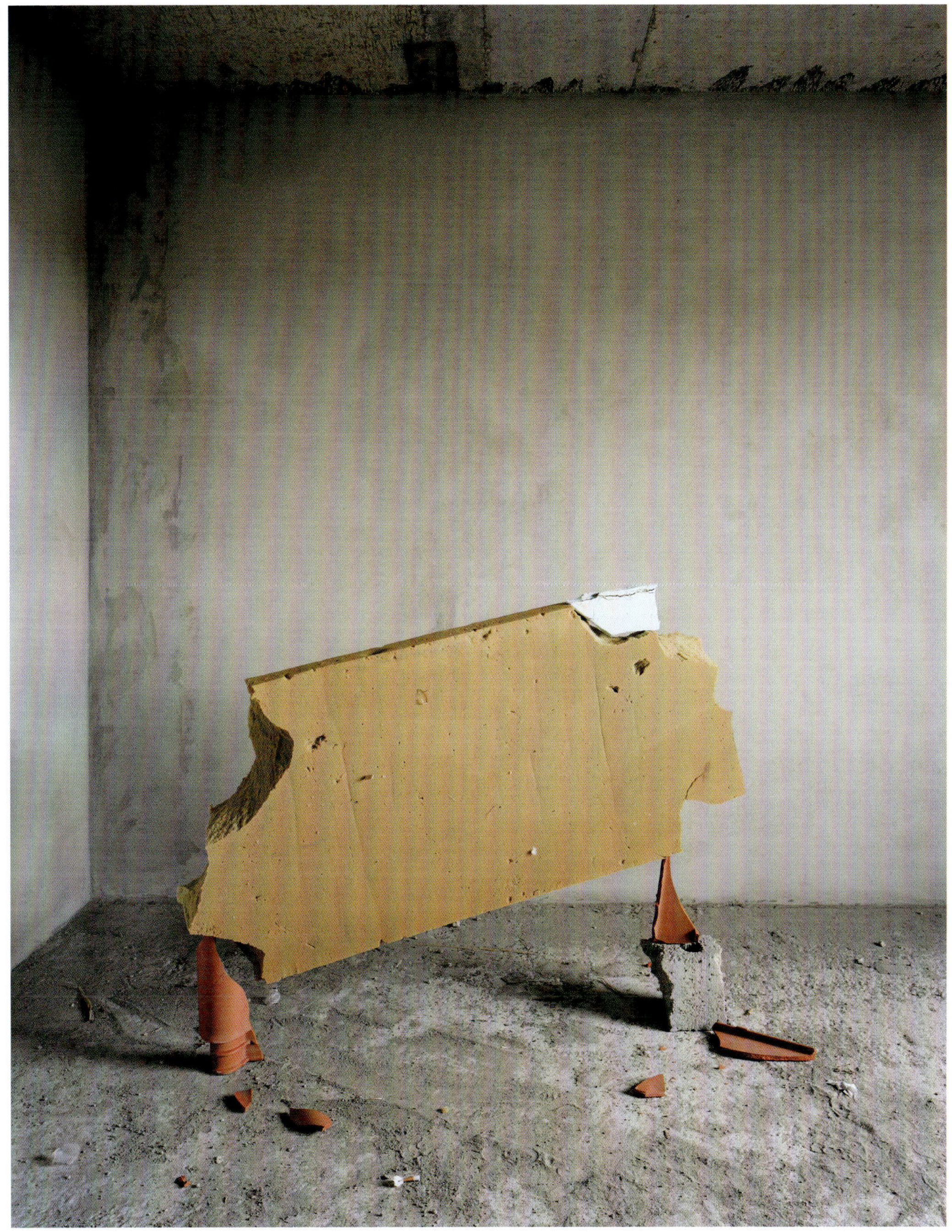

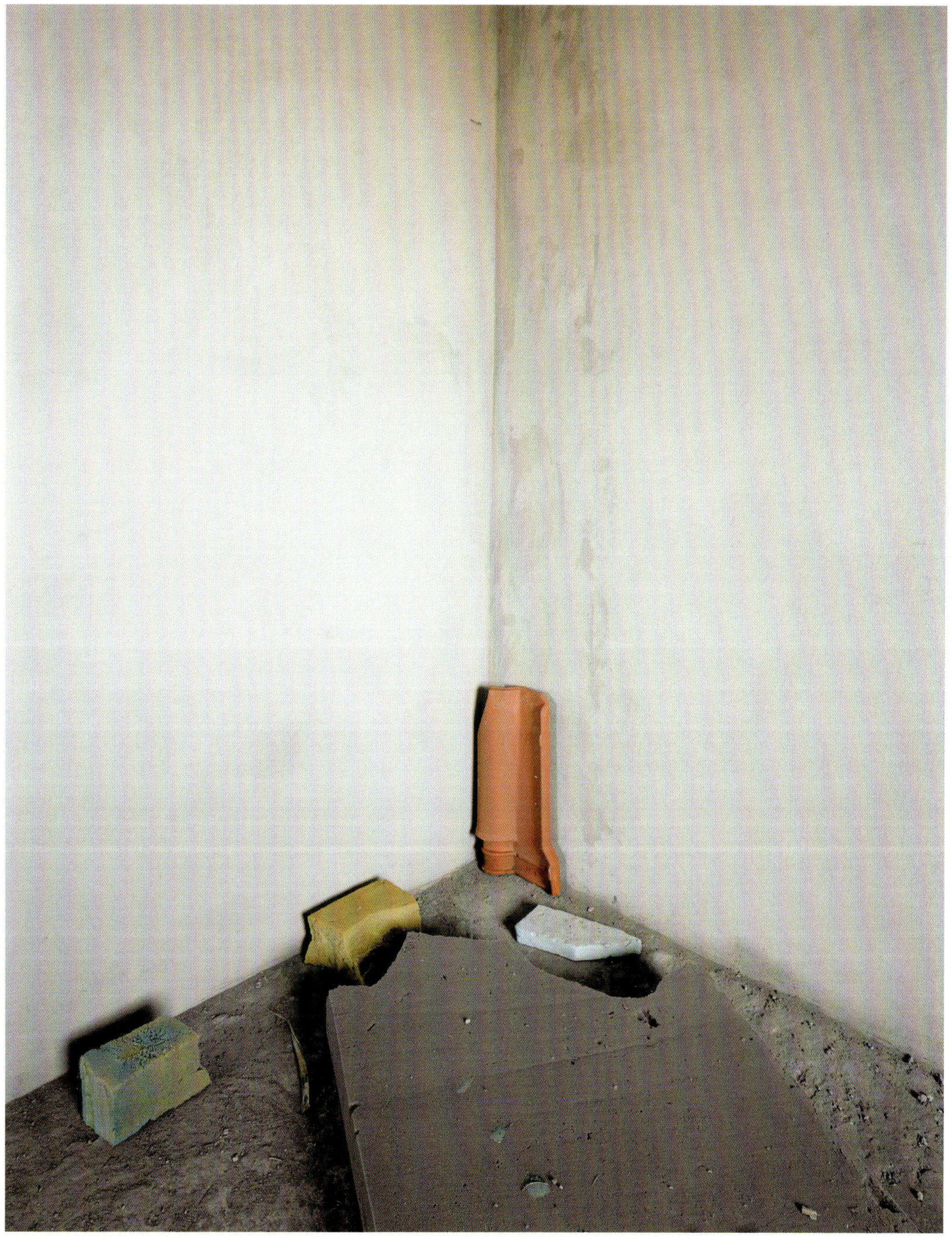

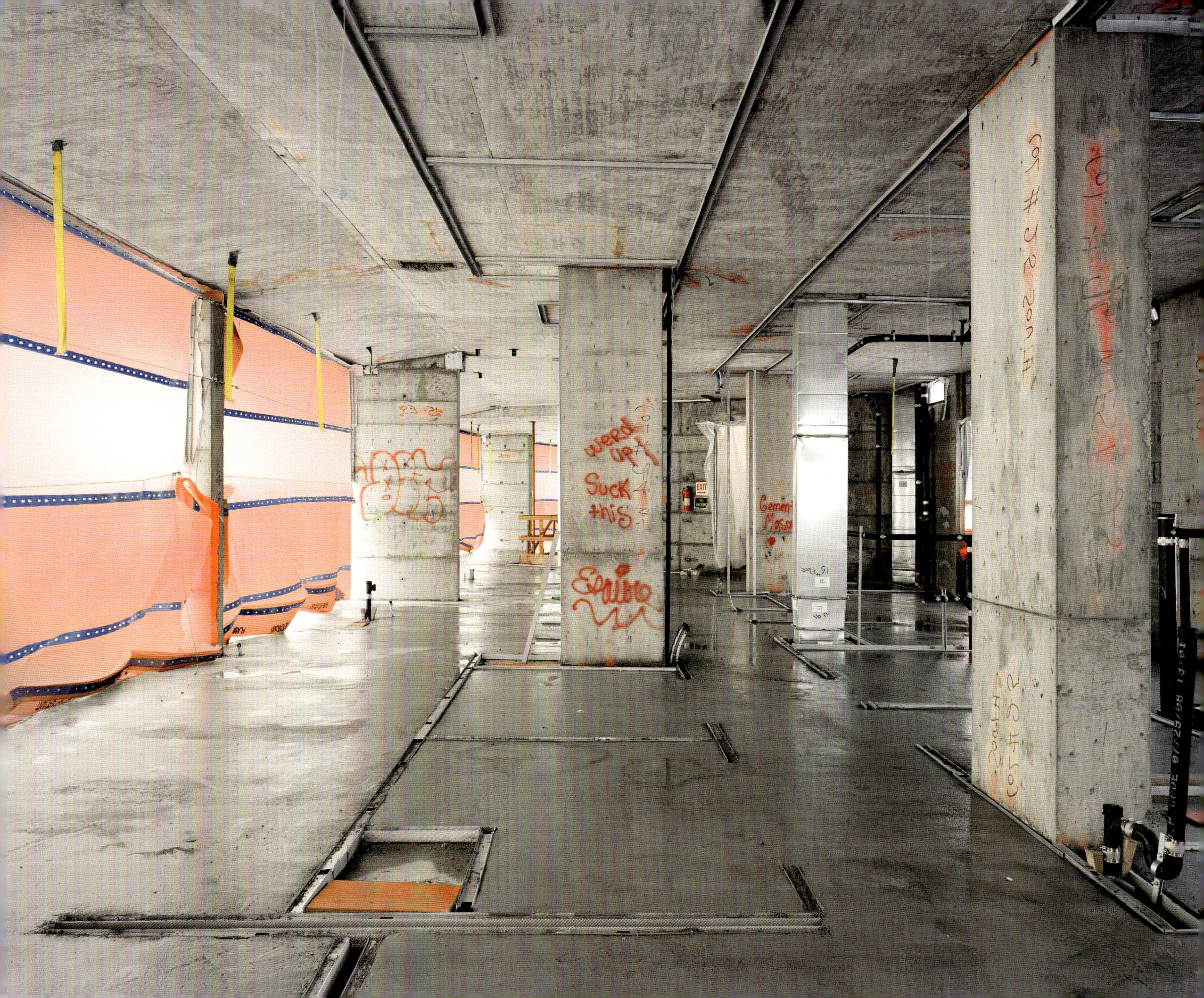
weed up
Suck this
EXIT

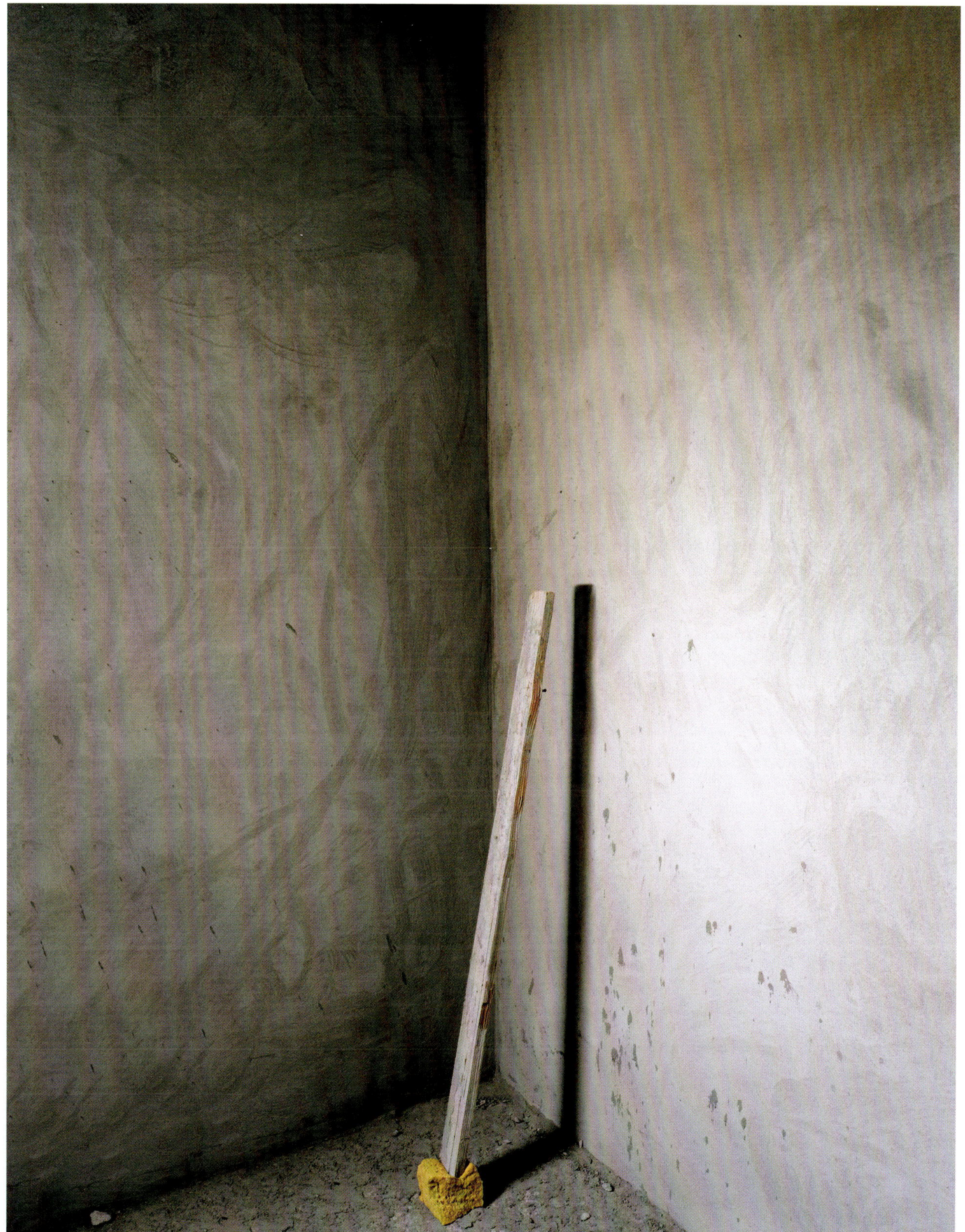

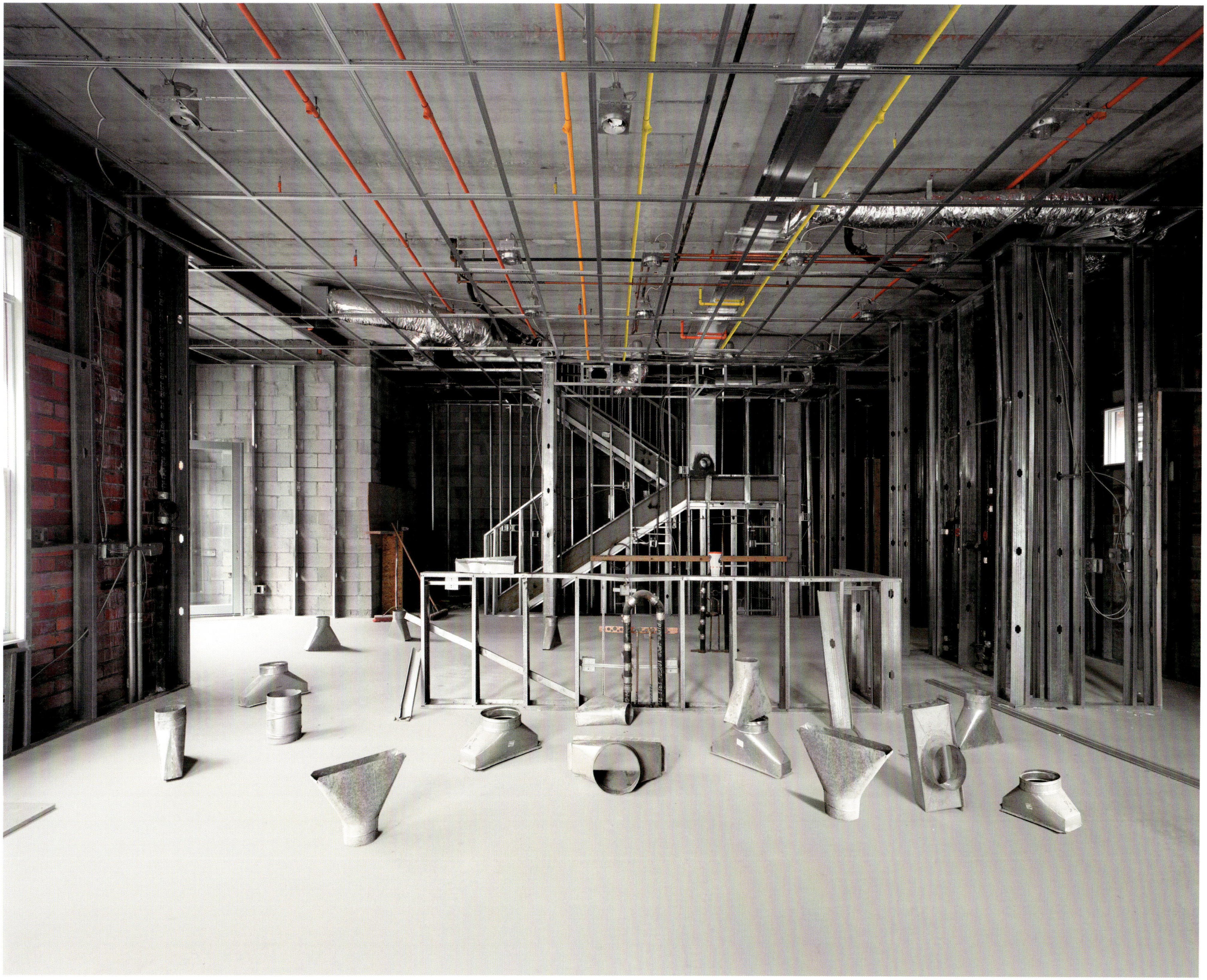

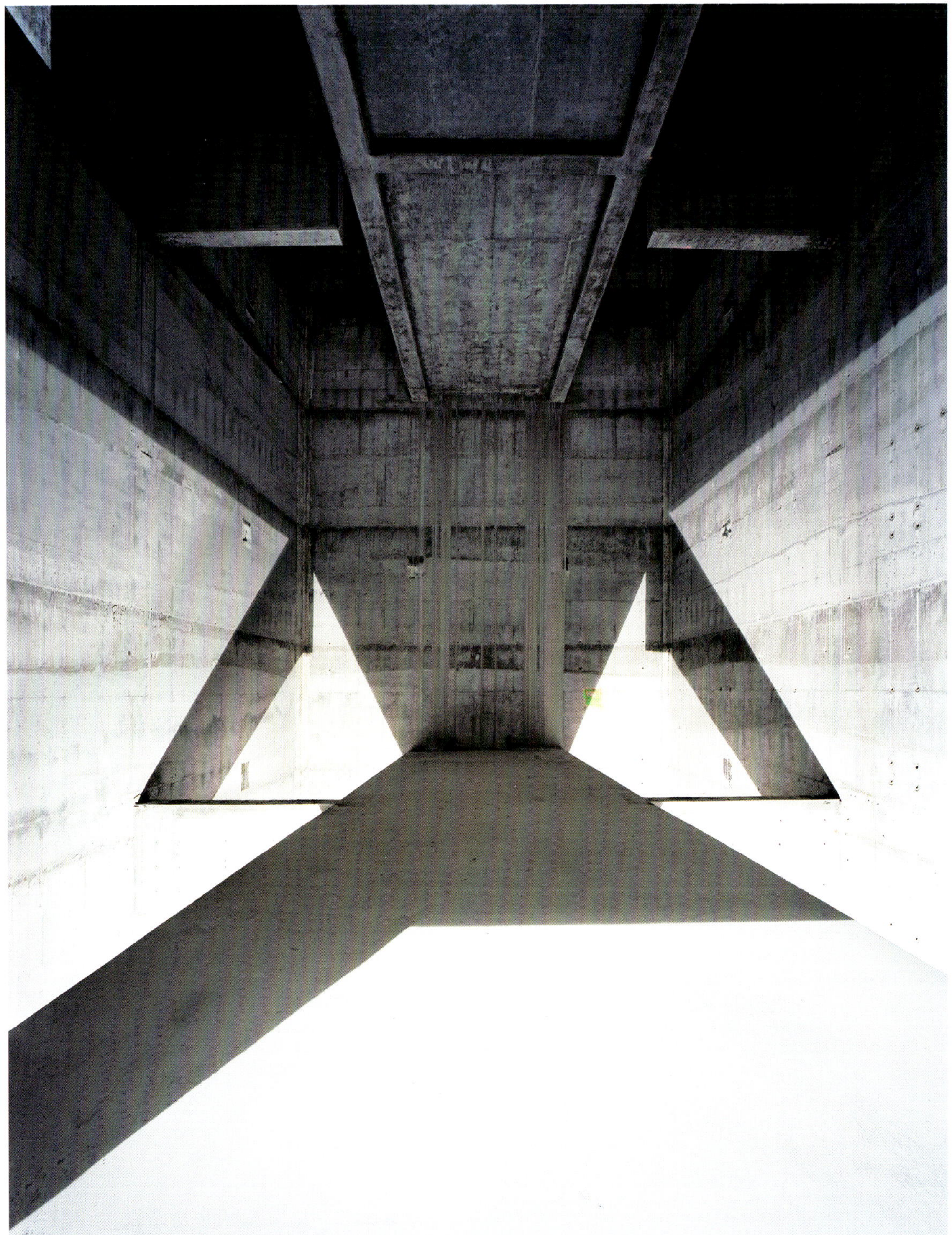

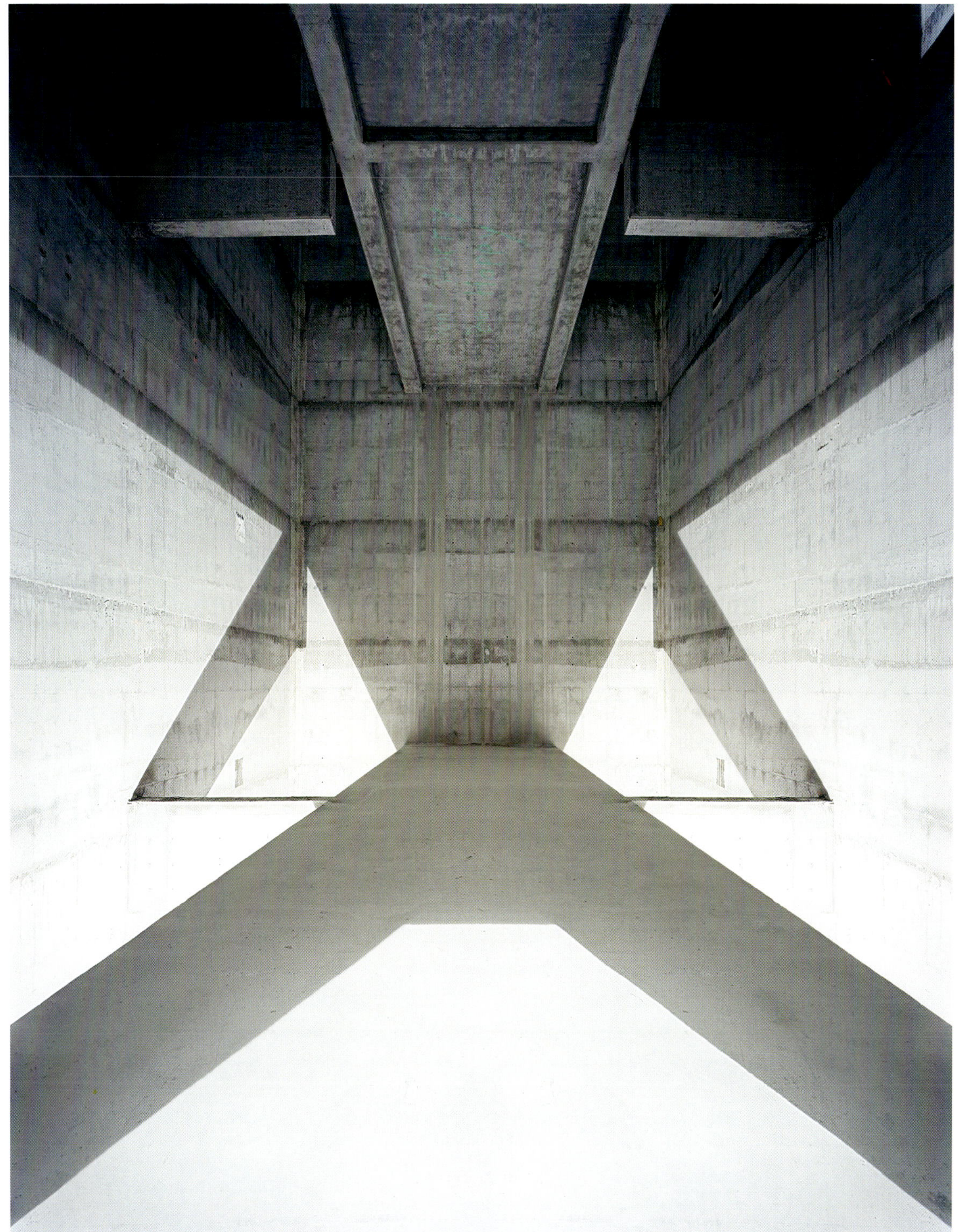

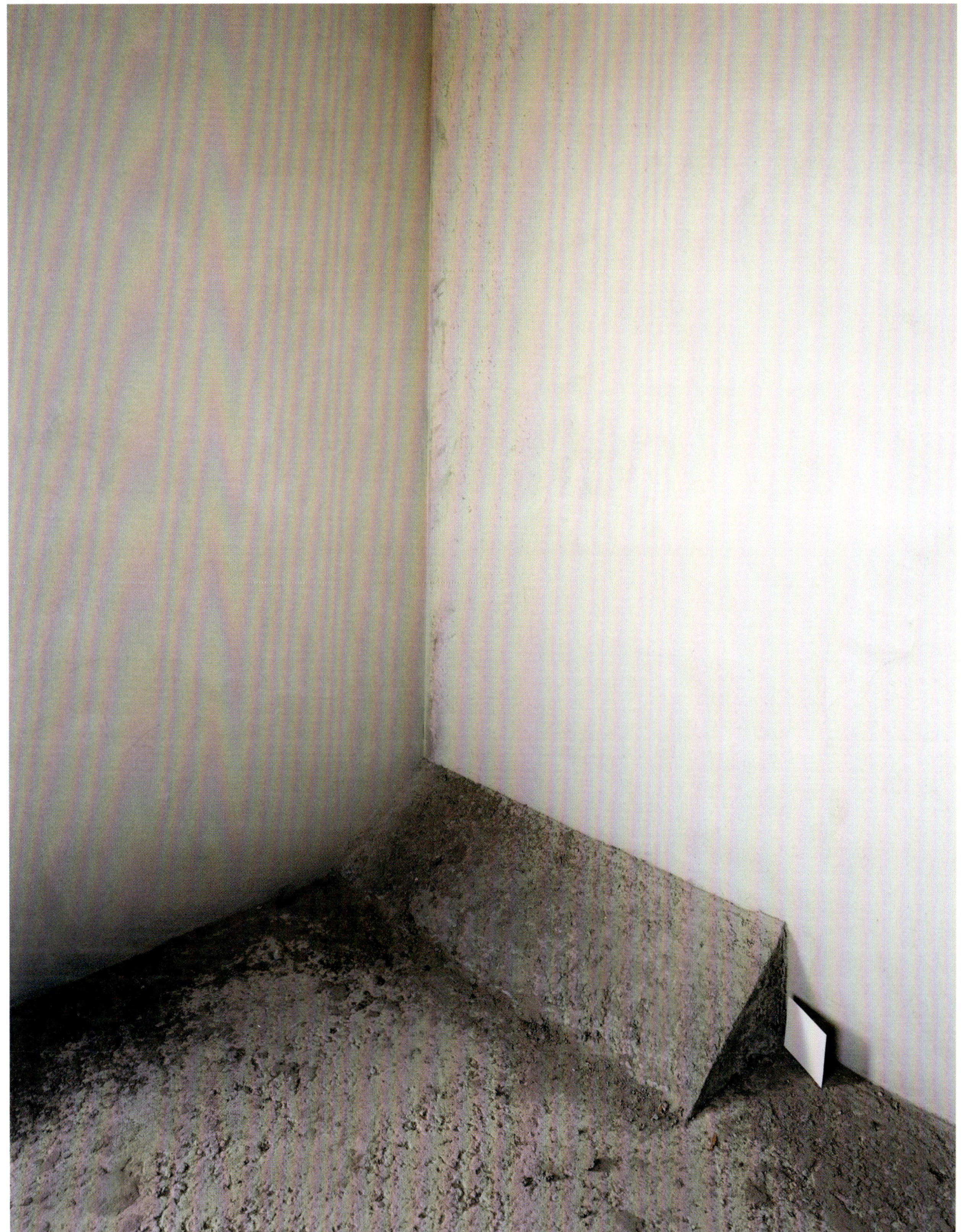

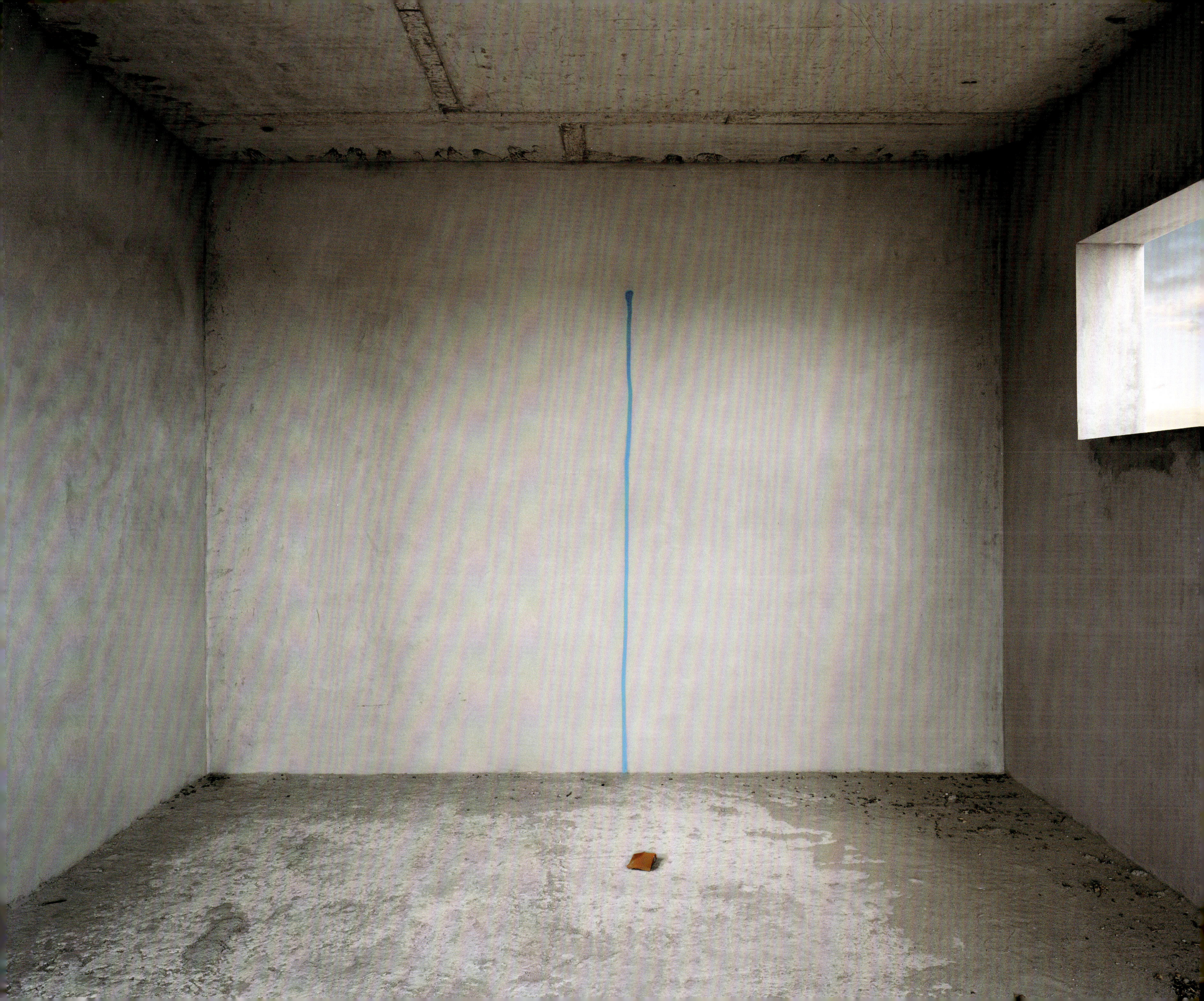

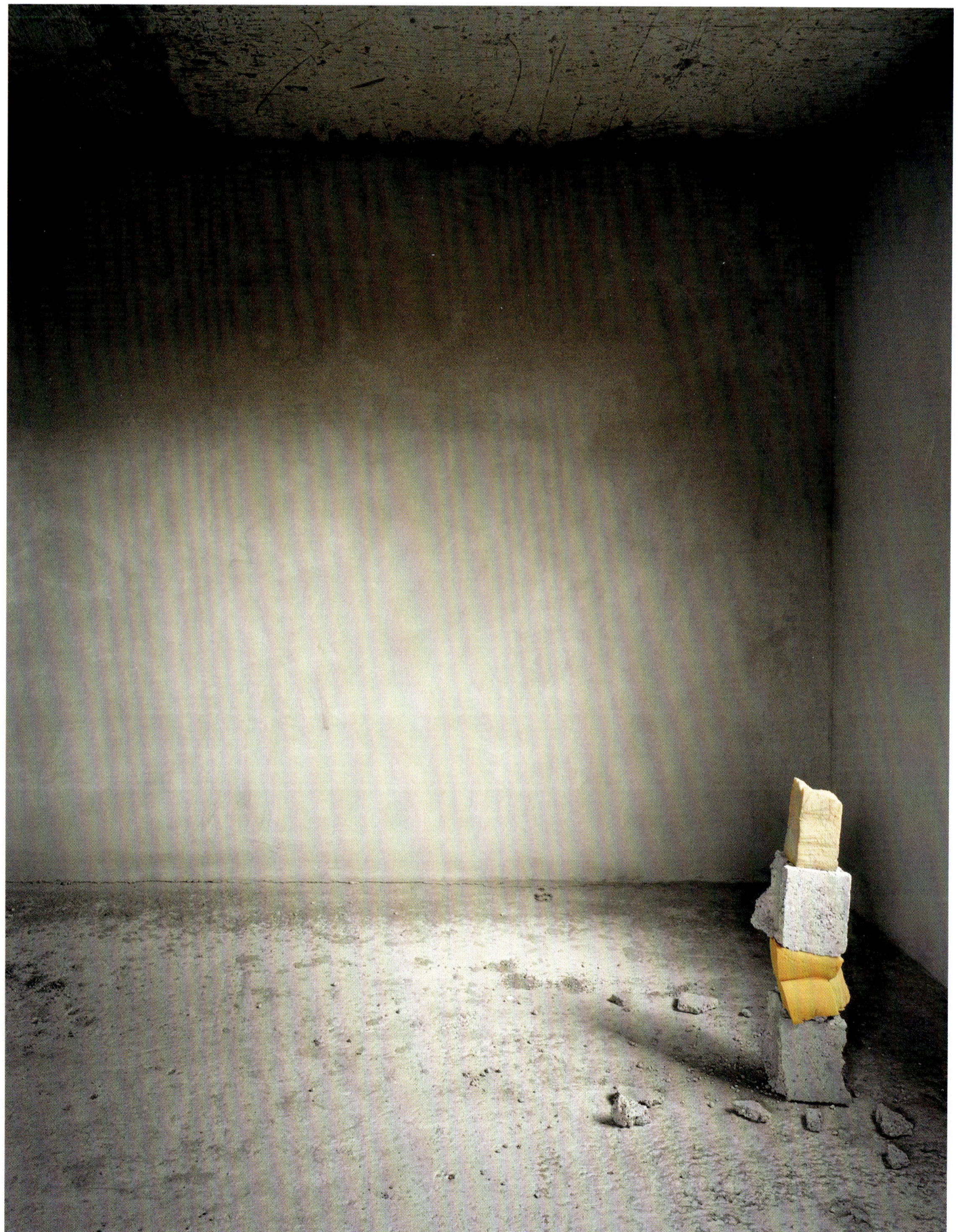

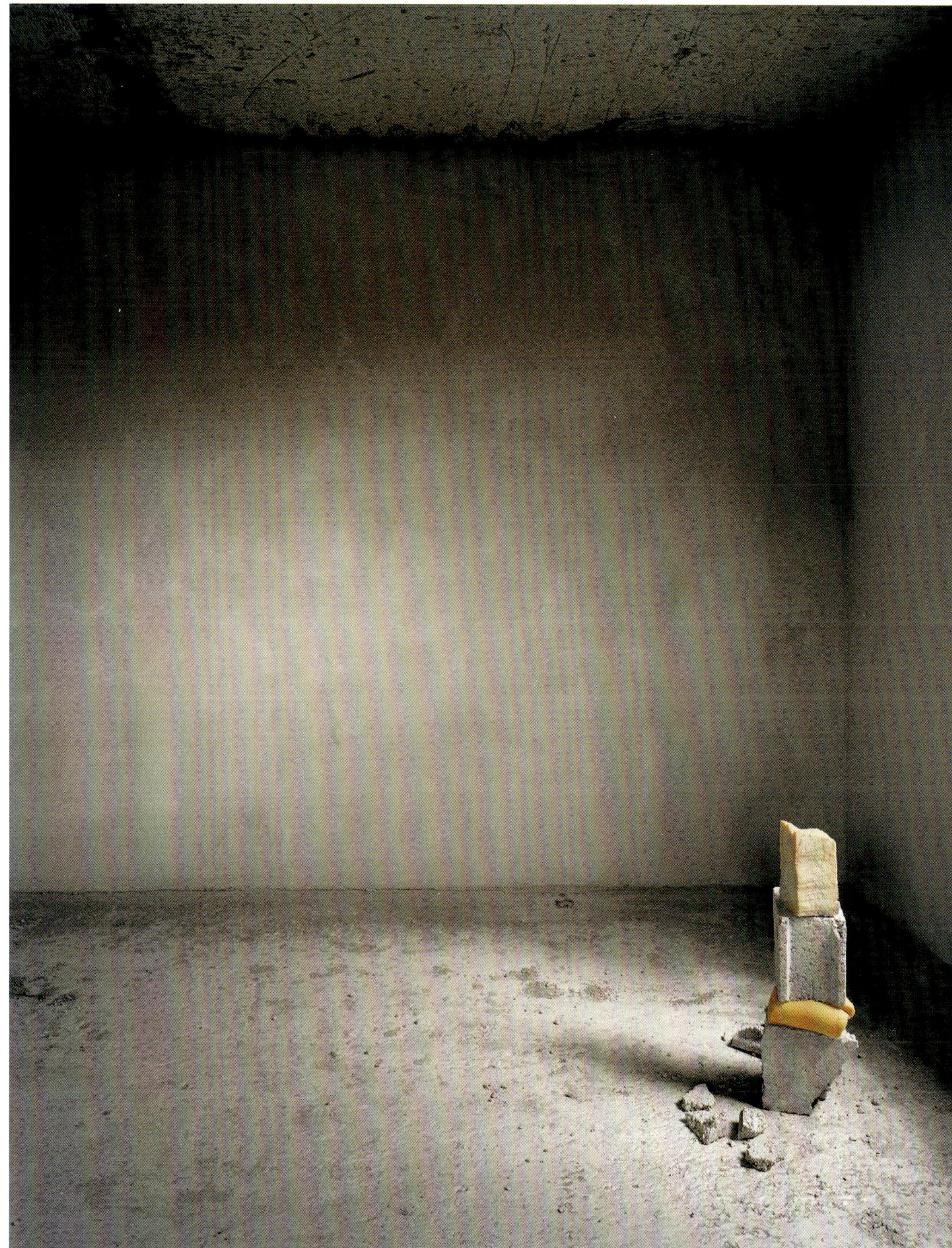

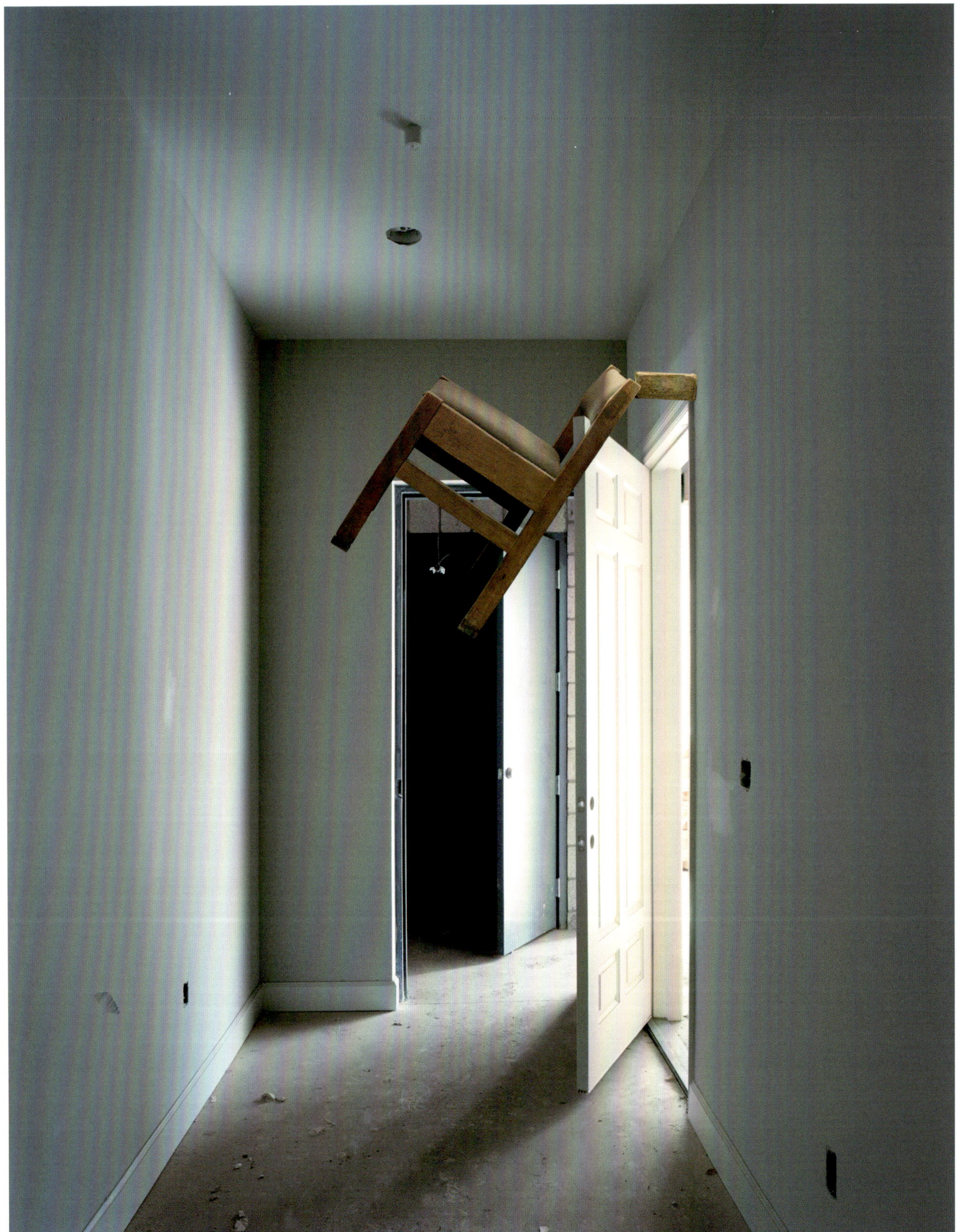

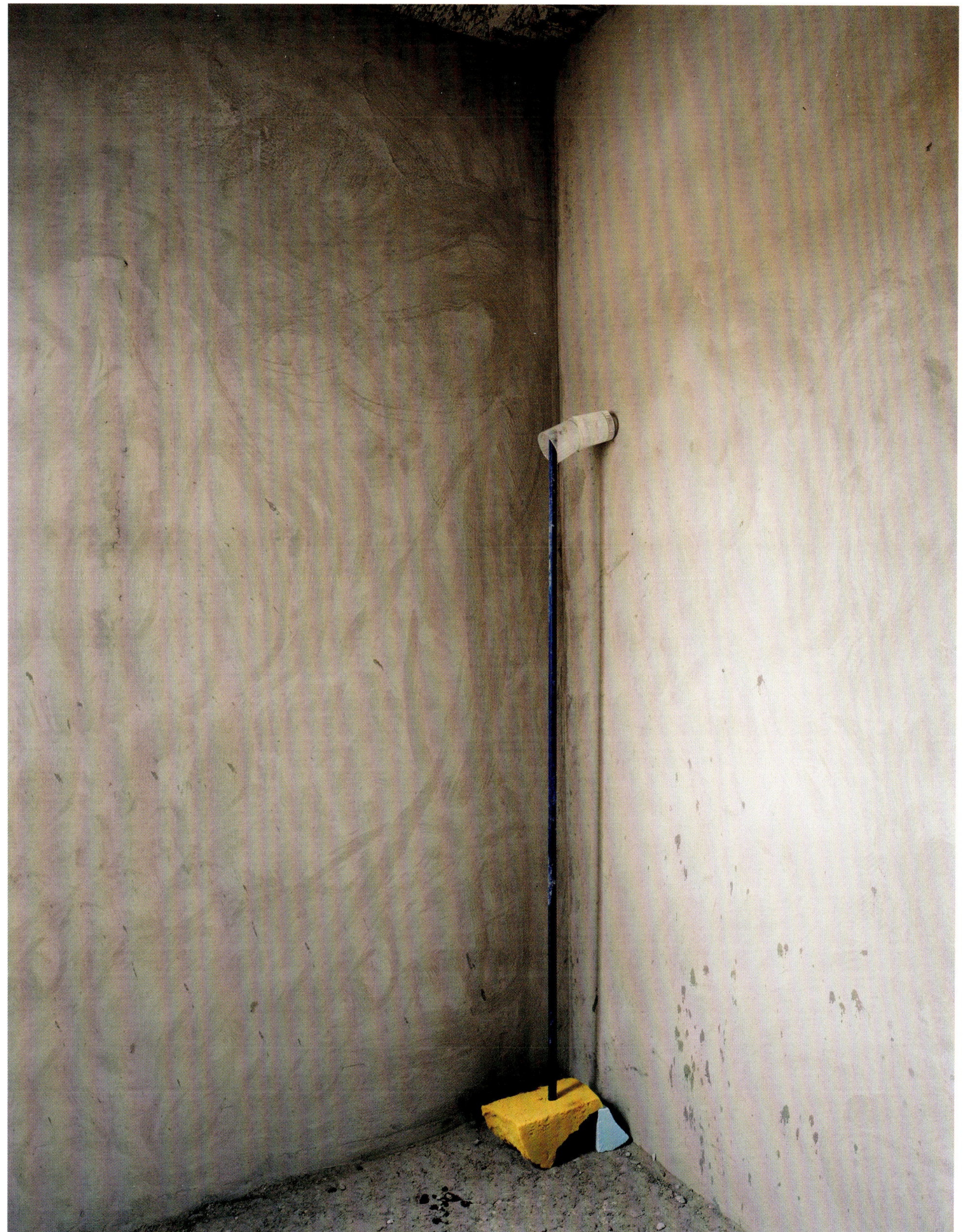

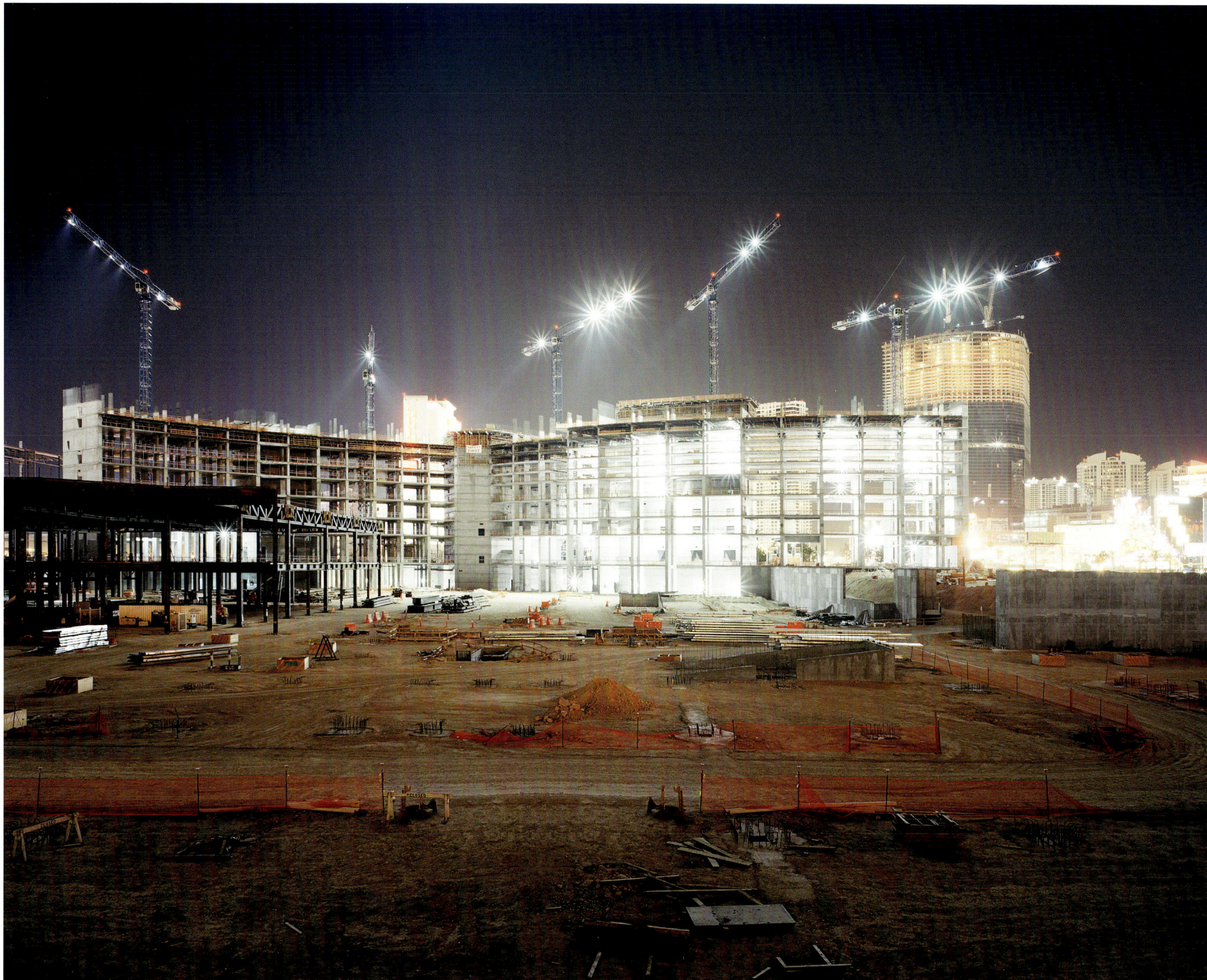

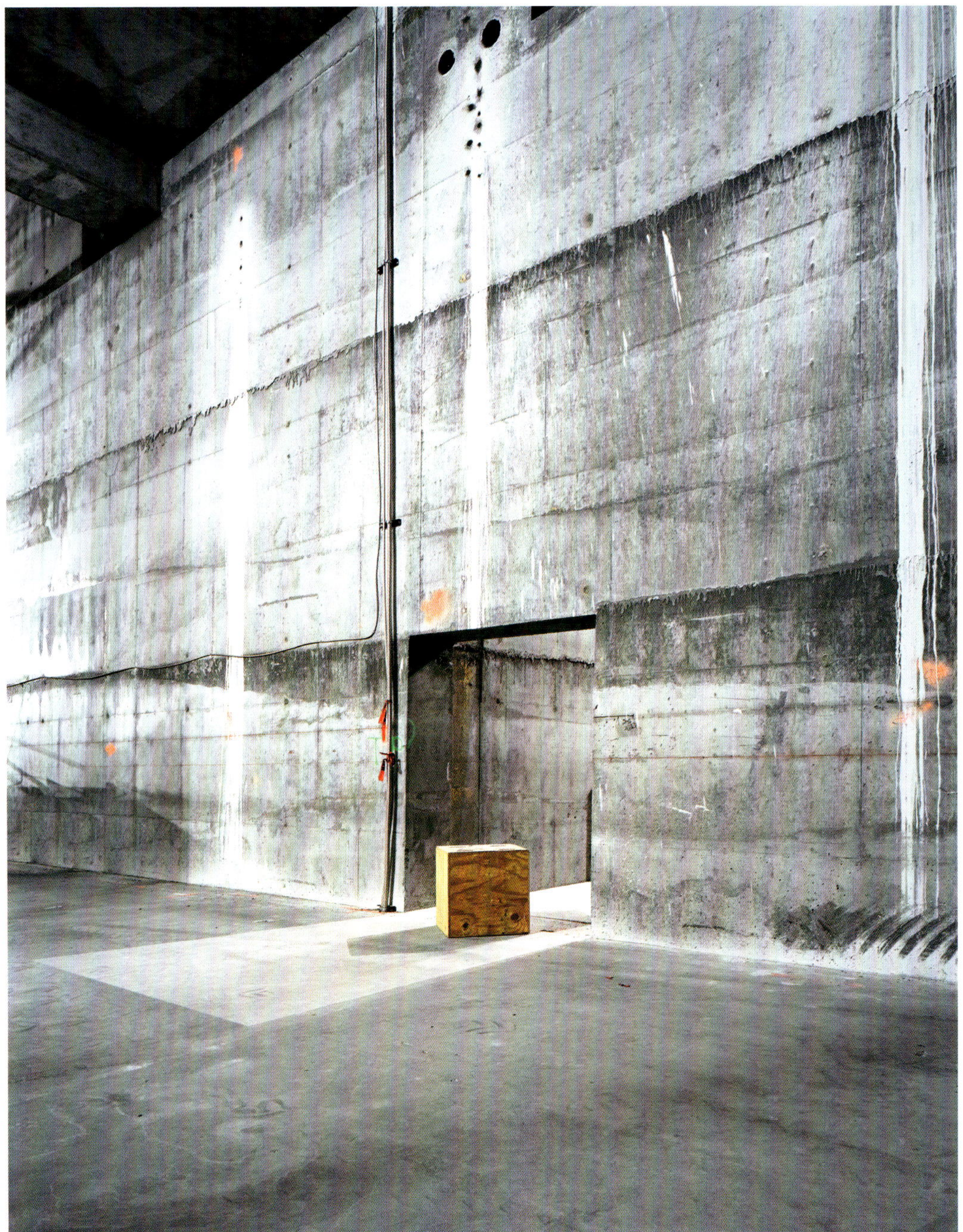

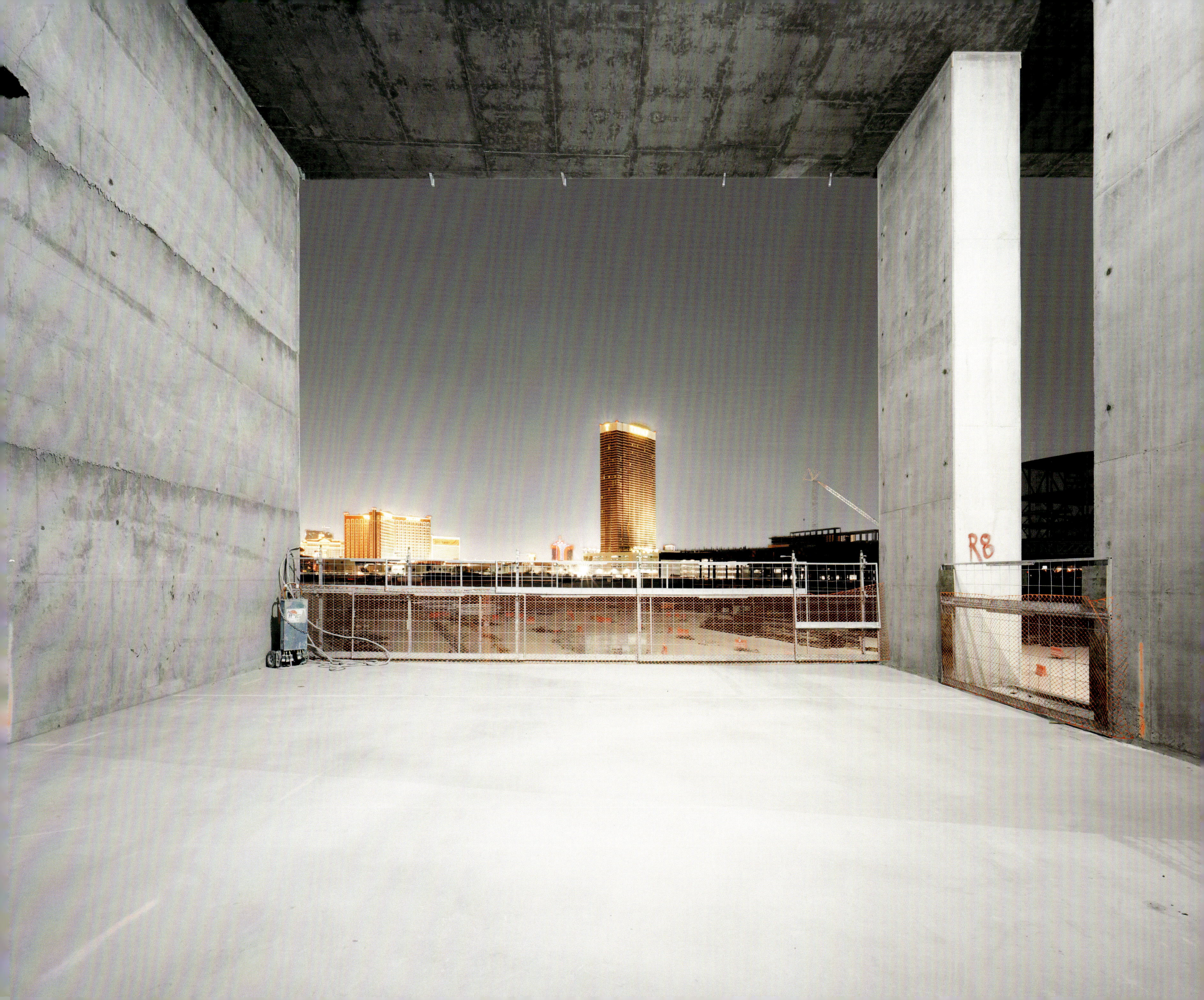
R8

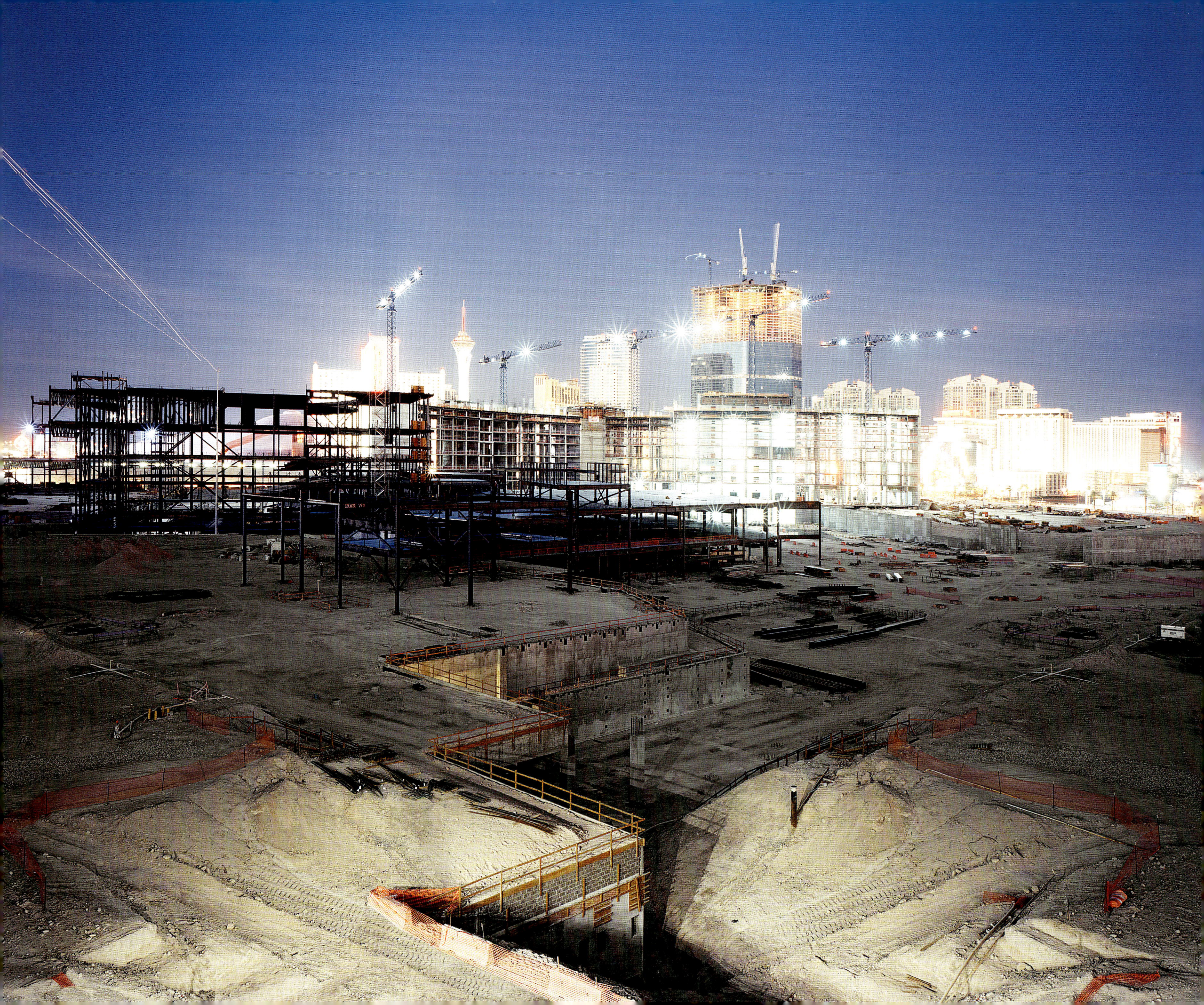

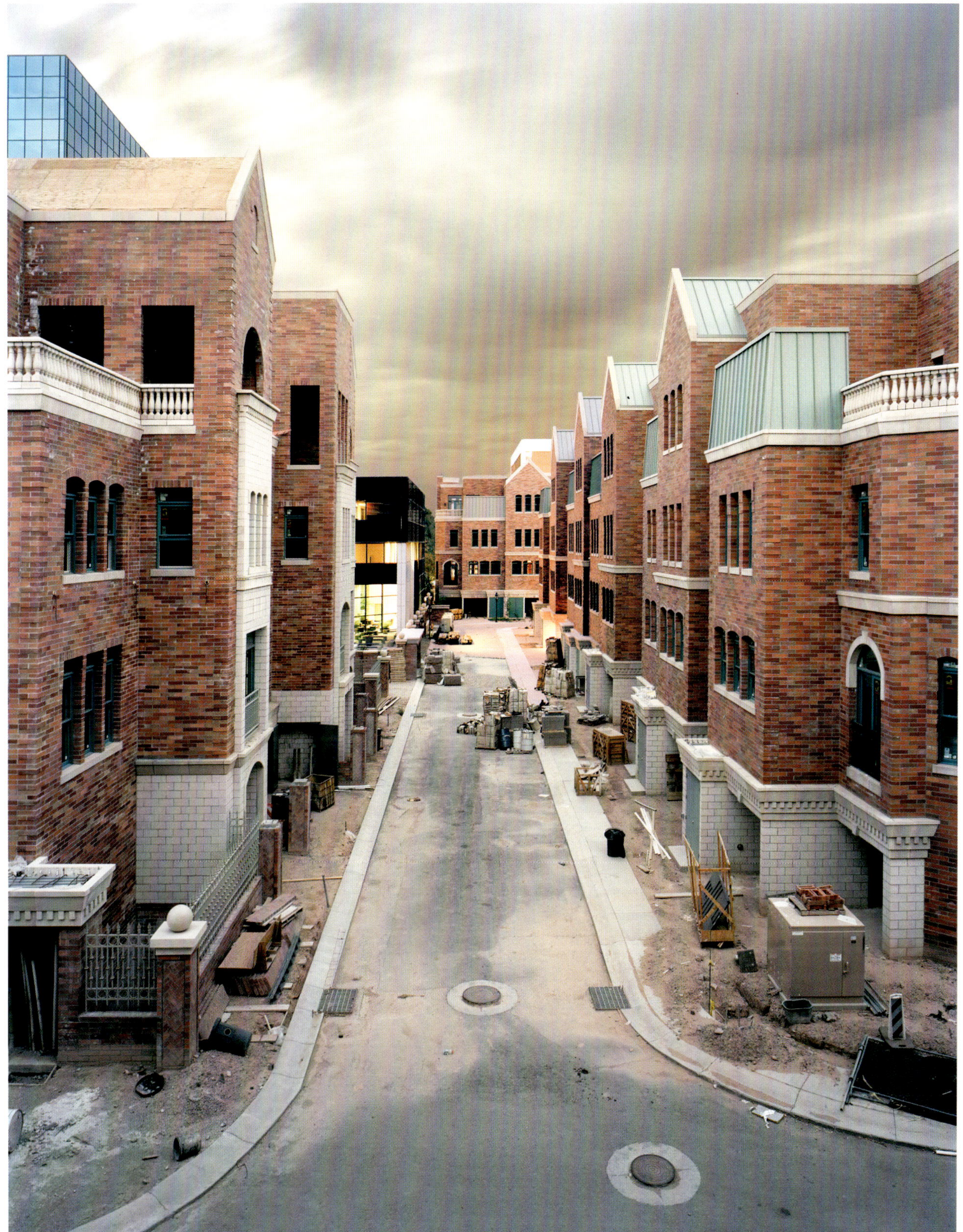

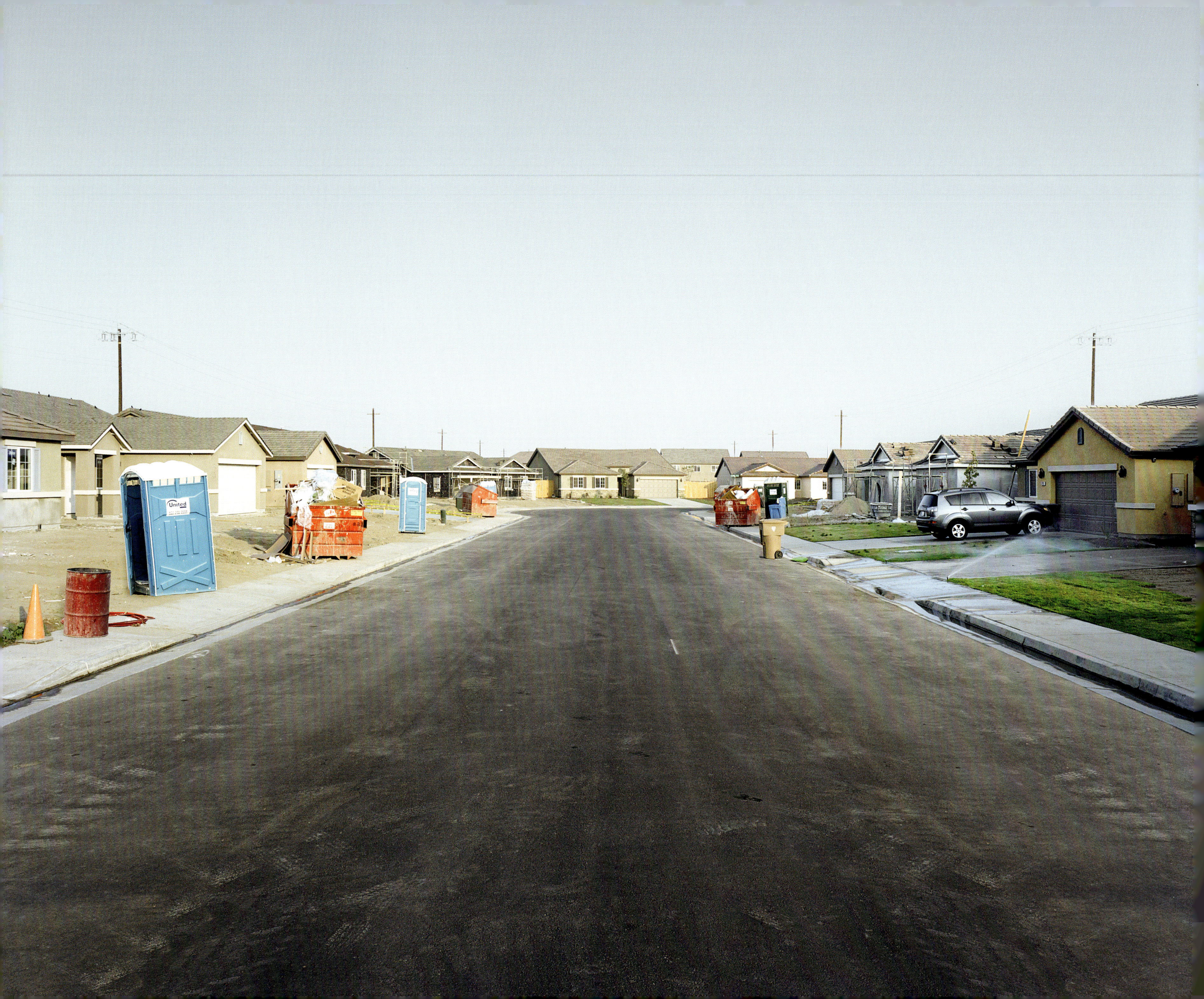
United

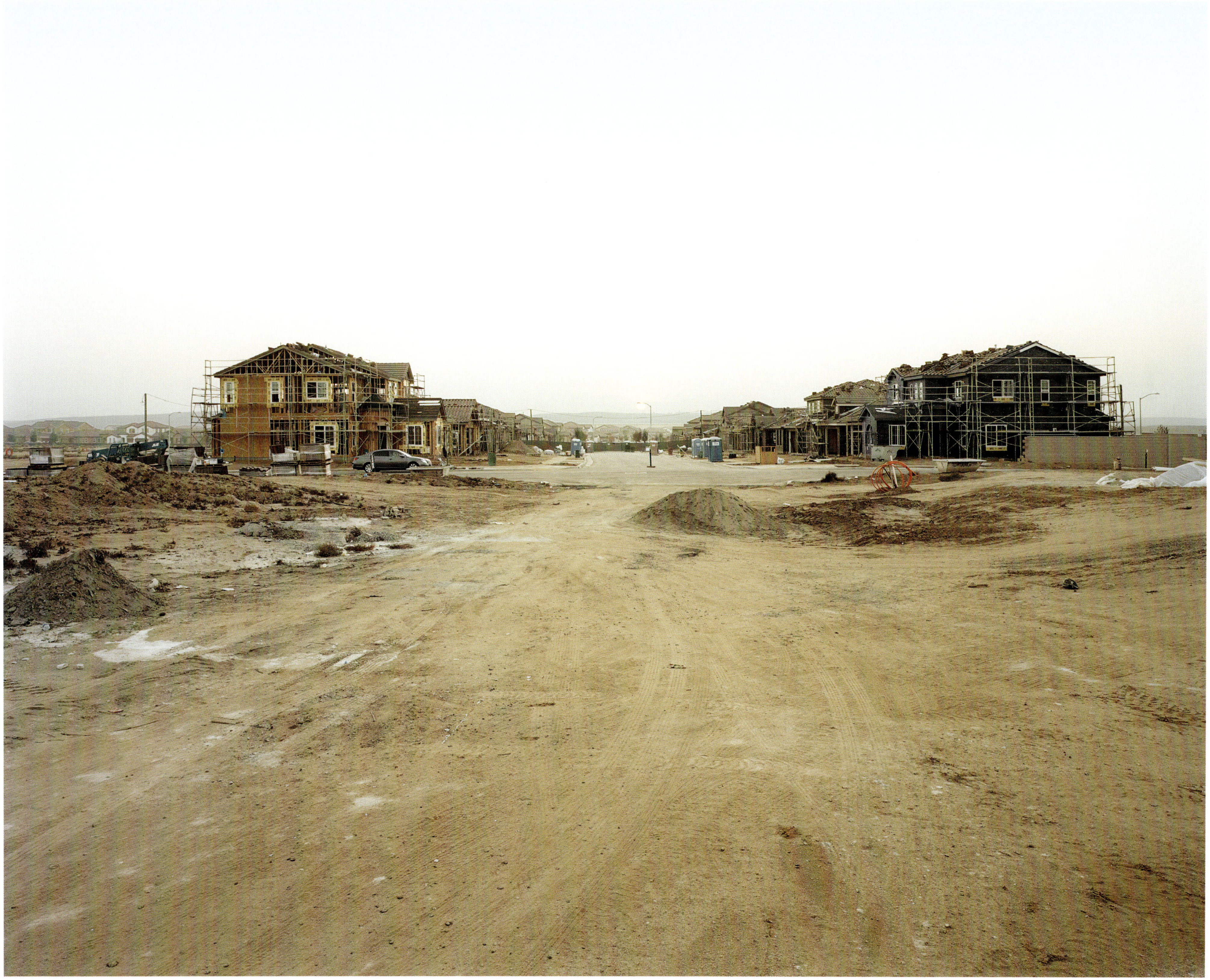

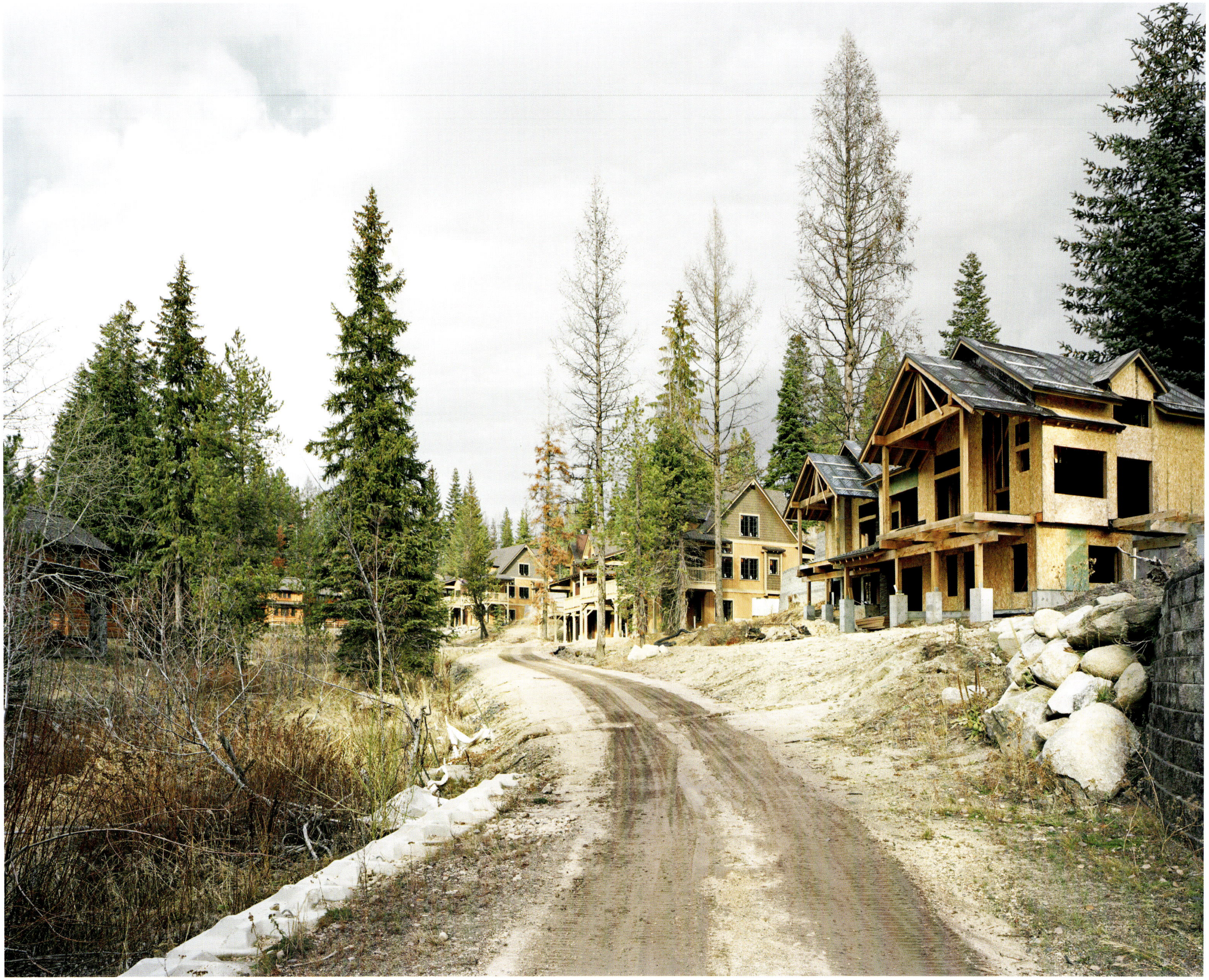

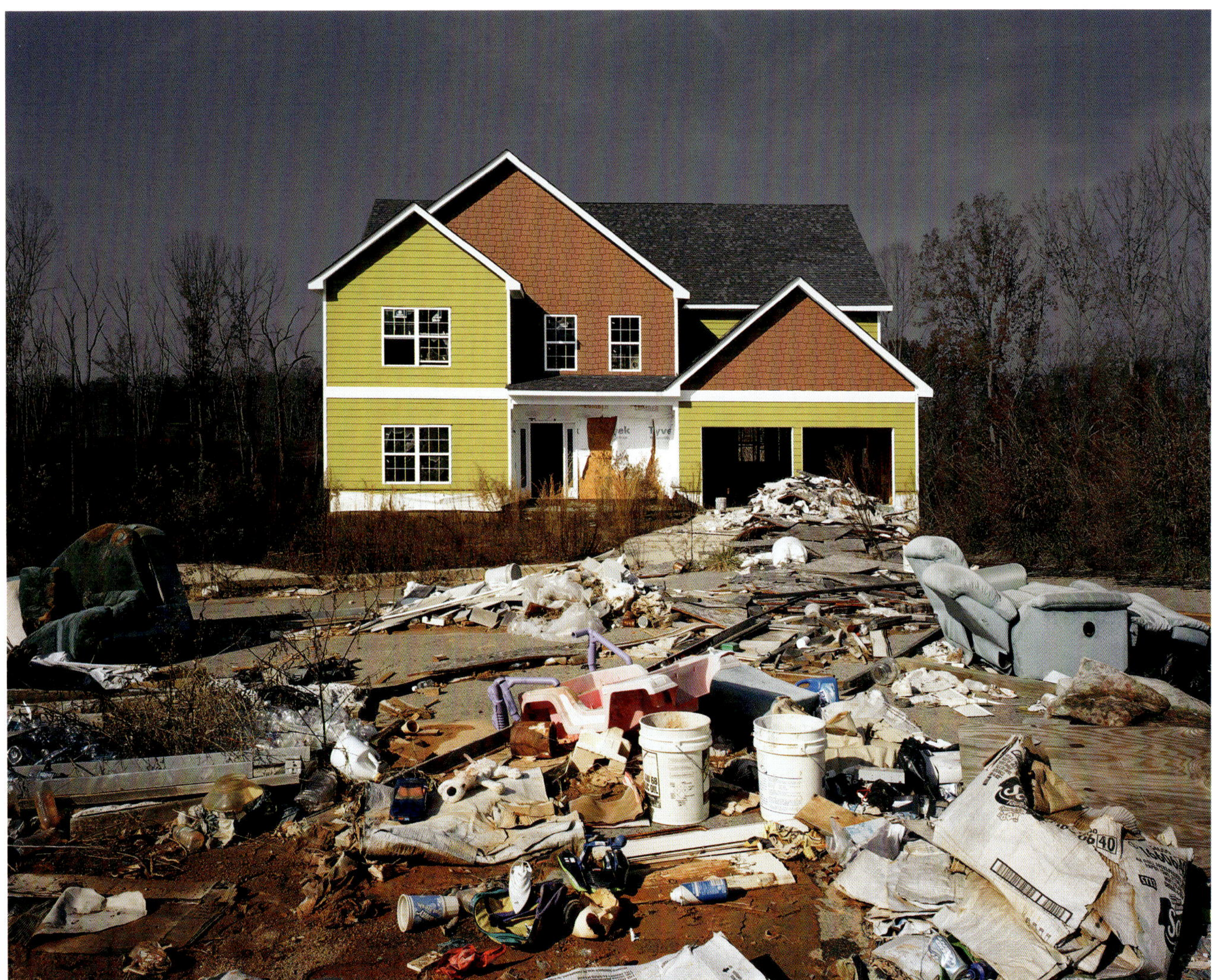

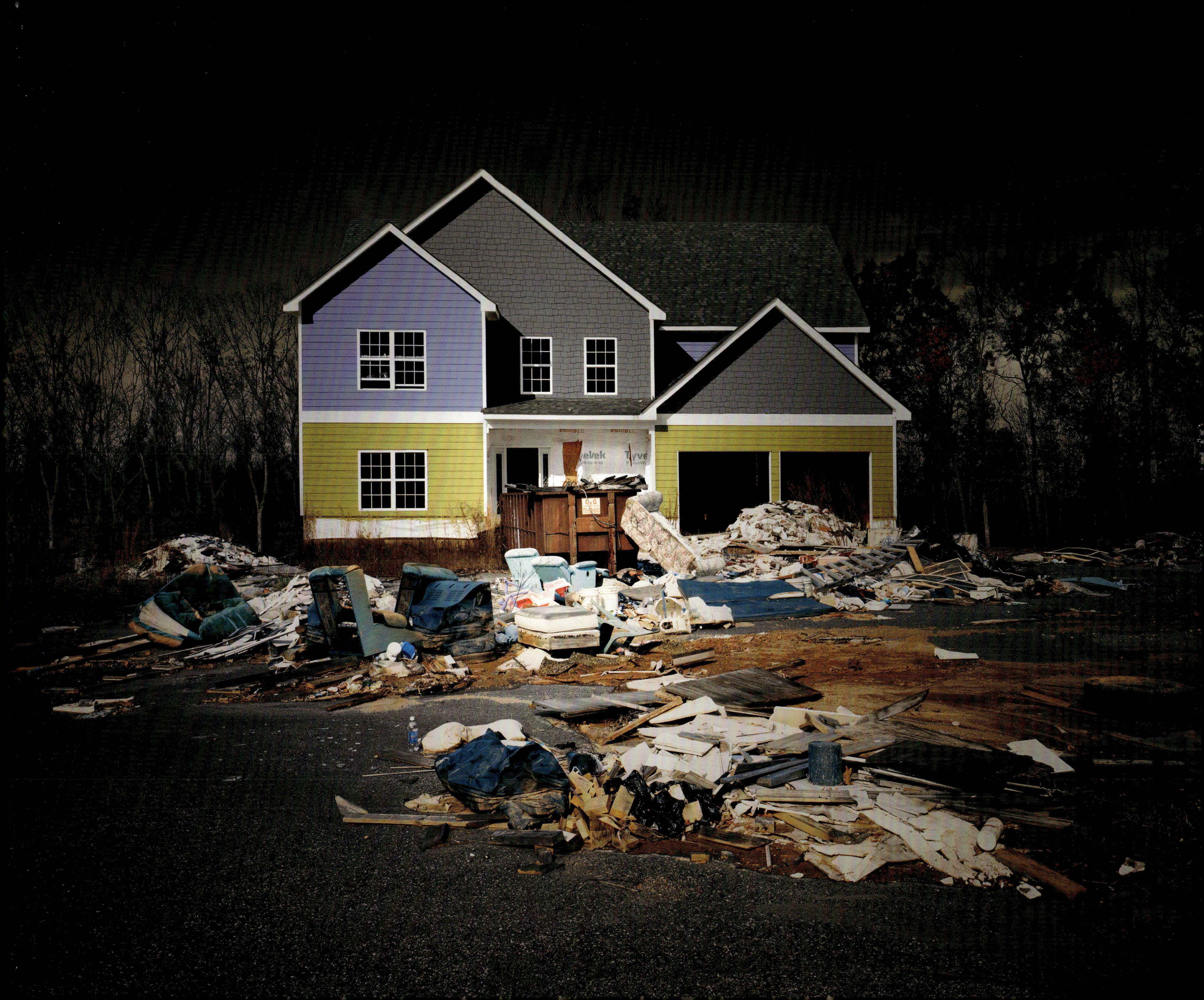

PETER D.OSBORNE THIS IS NOT A HOUSE

1

The Greek word *Oikonomos* (economy) derives in part from *Oikos*, meaning house. This is still present in modern Greek, which uses *spiti* for house, as well as related terms such as *oika*, *katoika*, and *oikiakos*.

2

...Or through the windows we shall see
The nakedness and vacancy
Of the dark deserted house.
Alfred Lord Tennyson, *The Deserted House*.

To begin let's take just one image (next page). It is a frontal view of part of an American house, a fragment, the corner of the house's face brightly illuminated. It is a white house, a wooden clapboard house, probably in white pine, but it could be spruce, or cedar or cypress – all are utilised in the United States. The house has two storeys. The window on the second floor is divided into four wooden sub-frames that are in turn divided into six small glass panes. On the first or ground floor the window frame rises about a quarter higher than that on the second and is divided again into four smaller frames each further divided in this case into nine small glass panes. Some care has been taken with ratios. The two storeys are divided by a pitched half-roof sealed by grey tiling. It echoes the main roof glimpsed above. It is probably a goodly sized house to the average European and for millions of American citizens too. The mortgage on such a property would have been substantial if not astronomical. But the house is not distinctive. Some rendering of the white clapboard style is pretty well America's default domestic architecture. It is classical American. It goes back to the settler beginnings of Anglo-America. It references an established tradition, that of deep (Euro) Americanness. It proclaims belonging – to home and nation, and the affluent middle class.

Of course we can see that this home has been undermined. The windowpanes have been shattered. Jagged glass shards remain as indices of some kind of assault. An unease in the relative proportions of the building has prepared us for this. Like a wide collar, the half-roof extends dramatically. It exceeds the extension of the main roof by an order of three. While it would have afforded increased shelter it unbalances the overall design, undermines what symmetry is striven for. And on closer inspection it is clear that the frontage has never been developed. The house is damaged and yet pristine. Perhaps it has never been inhabited; the client withdrawn, the builders departed before completion – a ruin before it ever became a dwelling.

But this is not a house; it is a photograph; and present in the photographic qualities is an element more disturbing than these signs of incipient destruction. The white forms of the building stand out against the darkness within, around and beyond it. It is a blackness both profound and absolutely flat, actual and equally abstract. The house

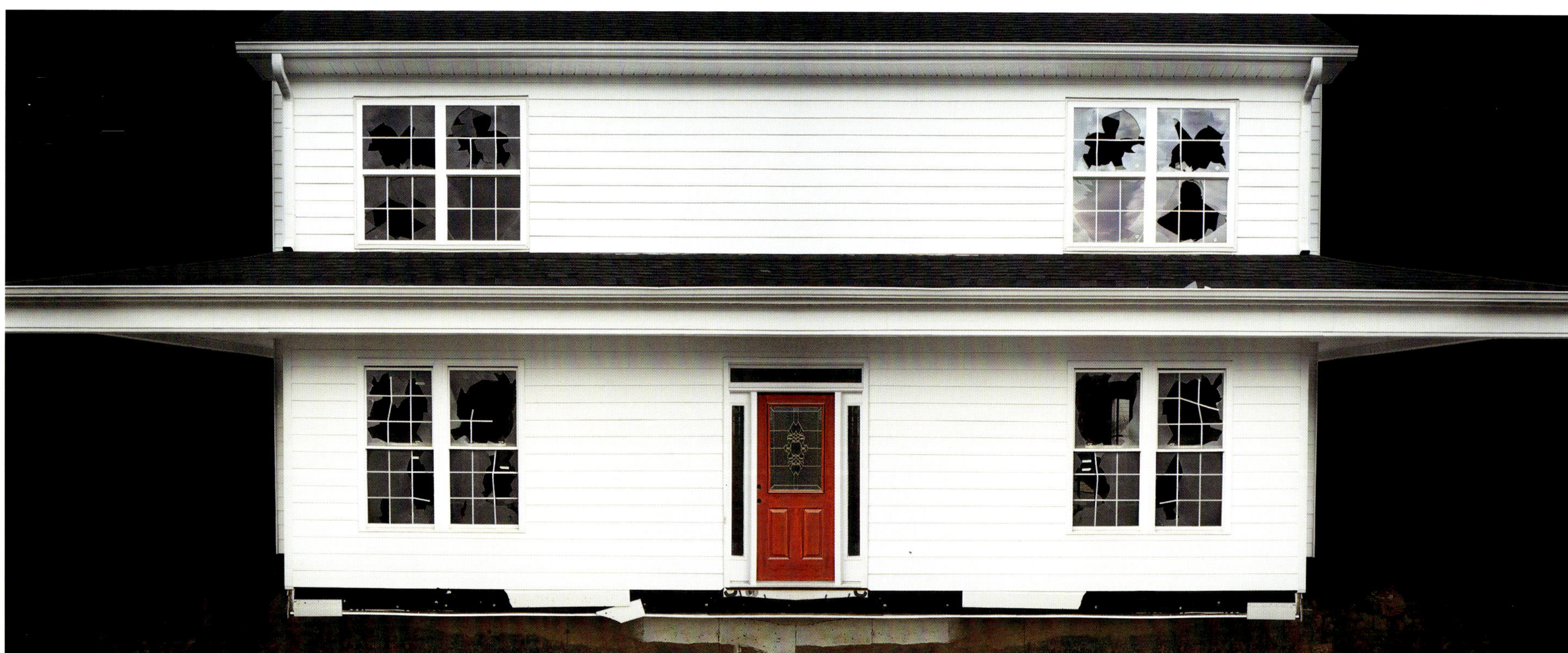

contains this darkness and is contained *by* it. Is this just the darkness of night? Or is it an aesthetic device – the black against which to throw the white forms into dramatic relief, presenting the uninhabited house as pure design, utility become beauty? Or is it the mark of an absence that haunts all these houses and of the forces of negation that produced it – economic, political – even metaphysical?

3

A house constitutes a body of images that give mankind proofs or illusions of stability (…) Without it, man would be a dispersed being (…) It is body and soul. It is the human being's first world.
Gaston Bachelard, *The Poetics of Space*

Ruins remind us of the mute materiality of the world, of the stuff that cares nothing for our futile projects, that recognises nothing of the order our architectures try to impose on the world. Yet, ruins are for the most part one of history's set designs and from out of them meanings are erected, representations made. Think only of the photography of American ruins: the burnt buildings of Civil War Atlanta in 1864; Arnold Genthe's vistas of San Francisco after the 1906 earthquake and fire, or Mark Klett's photographic reflections on the same event from the city of 2006; the abandoned sharecropper's shacks in the FSA photographs; the photography of the ghost town; Clarence John Laughlin's deserted and decaying Mississippi plantation mansions shot in the 1940s; the press photos of the torched tenements of Newark and elsewhere during Black America's risings of the 1960s; Robert Adam's provisional-looking Tract house projects in the desert; Joel Sternfeld's dream house undermined by landslide, or his sites polluted by atrocity, places of ruined memory; and of course, there is the Ground Zero imagery of Meyerowitz and countless others. American disasters tend to become iconic. There are strings of websites devoted to the photography of abandoned motels and of ghost towns. One of the images from this work has already been appropriated by one as an example of 'the new ghosts towns' of 'subdivisions and MacMansions' that people can no longer afford (www.inquisitr.com/28369/the-new-ghost-towns/).

The history and imagery of the ruined shelter or settlement are required to furnish the continuing epic of travelling and moving on that is the USA – or do we now say, 'was'? – an epic of disasters endured and overcome for sure, but one ever attended by a sense of America's precariousness, of its own brevity, its own uncertainty about where and if it belongs. In this regard James Wines' cracked and crumbling Houston BEST store built as a ruin, as though already shaken apart by tremors, is either a snoop cocked at fate or an inoculation against it; or perhaps a claim to permanence posing as its opposite, that is solidified ideology.

European America, then, is a Settler Nation, and at times an unsettled one. The house or the shelter has special connotations there that convey the story of established communities or of the mass triumph of privatised middle-class family life initiated from the 1890s by Sears and Roebuck catalogues and stores. At the same time, from the cabin-in-the-clearing, through the circled wagons to the 'gated communities', the house and shelter is stalked by an accompanying history of fears and beleaguerment. Any disaster concerning the shelter, the settlement, is extended very quickly into a metaphor for a whole historical process.

America is also the embodiment of that precarious project Modernity, and Modernity, argues Henri Lefebvre, produces two contradictory conditions. On the one hand, there is the promise of 'Comfort', of which the Home is the core both ideologically and in terms of material well-being. On the other, says Lefebvre, lies the continuous imminence of what he calls 'Ferocity', which can appear in the shapes of war, terrorism, technological and ecological disaster, and the visitations of economic crisis or collapse (Lefebvre, 1995: 190-191). Martins' images present us with the points at which these conditions have recently collided. As Norfolk, Ristelhueber and others developed the 'Late Photography' of the battlefields of the Gulf and Afghanistan, so Martins has done so for the aftermath of the Credit Crunch (see Campany, 2003).

4

Photographs, says Vilhém Flusser, "dam up history in order to make it into a tableau"; in them, the scene replaces the event (Flusser, 2002: 128). Some of Martins' picturings in this work do evoke the idea of staging. But the effect is more than simply theatrical, a deferment or denial of history, for it is the historical or social reality of the 'toxic economy' that remains dramatised in the images as a condition of seeing, or rather of the visibility of the object. If the photograph halts the narrative of history here the effect is relevant, for depicted here is development paralysed, a disabled project – stasis, that condition so feared by modernity. The buildings stand before the viewer like the scenery of a bankrupt theatre, the sets of an abandoned movie project. Nothing moves in or through them. It would be a mistake to assume, as some appear to have done, that the visual elegance, the abstraction, the careful rendering of formal values and the necessary manipulation that exist here as in much of Edgar Martins' practice are inappropriate qualities when applied to the themes of this present work. It is true that the crisis and human unhappiness indicated in many of these images is real enough, whatever the class of those who bear them, and they impose on any photographer some kind of ethical responsibility. However, Martins has never been a humanist photographer, nor is he a social documentarist. There is coolness, and a distancing throughout his work, and an overriding concern with form. There is even a case for describing him as a metaphysical photographer. And yet it is precisely certain of these qualities that seem so effective here. For example, in this work it is the absence of the human figure that pronounces the landscape human; the human is the principle that has gone missing, that has left a visual silence. This is a human crisis. And it is in the transmutation of lived spaces into near-abstract structures that the wider abstractions of the financial markets, of an uninhabitable economy, are indicated, made present in the deconstructions they have brought about. The departure of the human figure from these spaces is more than an aesthetic choice. Speaking through what Jacques Ranciére terms, "the silent language of things", these images depict more than an immediate actuality; they picture a *condition* which is social and empirical, yet which demands an aesthetics which cannot be served adequately by immediate observation or record alone; they by-pass both the dystopian melancholy to be found in some social documentary and the exclusion of the Social that characterises much of the art of voided places (Ranciére, 2006: 36).

In some instances the work summons up the conventions of landscape painting and photography, indeed landscape is the theme of much of Martins' practice. The representation of landscape is commonly based in the retrospective, distanced and even nostalgic viewpoint, often displaced from the present. Here its effects intensify the viewer's reading of the precise objects/events in the photographs. If one accepts Jean-Luc Nancy's reflections, landscape art has always depicted emptiness and loss. "Landscape begins with a notion, however vague and confused, of distancing and of a loss of sight, (*une perte de vue*), for both the physical eye and the eye of the mind." (Nancy 2005: 53)

Landscape, he says, comes into being when the human figure loses the foreground or disappears completely; when the gods have departed and the weight of allegory has lifted. It arrives as part of a deep transformation in the mentality of its time. "A presence", he writes, "is withdrawn", hence, "The landscape is the space of strangeness or estrangement (…)", (Nancy, 2005: 53 & 60). In Martins' landscapes it is less the Divine that has withdrawn than the elements of a *secular* sacred, that is, individual citizens and private property, inhabited wealth and, of course, beneath it all, the loss of Capital value. The frozen constructions and deserted interiors may turn out to be symptoms of another crisis to be cured. Alternatively, they may be signifying the relative decline of American power, even the destabilising of meaning or a system of meaning. Nancy notes that as the landscape form was ushered in by a loss or a displacement of meaning, it depicted place, "as the opening onto a taking place of the unknown", (Nancy, 2005: 59). No longer was land depicted as "location" (*endroit*) but as "dis-location" (*envers*), void of presence and giving, he writes, "no access to any elsewhere that is not itself 'here', in the angle opened onto a land occupied only with opening in itself" (Nancy, 2005: 59). The depth of absence in Martins' photographs is chilling: the half-made roads running among non-houses; the construction site fading into the twilight, seemingly losing all materiality; the golf course with no players; a remoteness seeping into the neighbourhoods that failed to come into being. Perhaps an even greater absence lies in the loss of our ability to name, to recognise these places. They are no longer nature as development has overtaken them; but they are not quite material culture either, the development has halted and gone having invented 'Nowhere', which, as it happens, is the meaning of More's word, 'Utopia'. These places, unnameable, suspended between categories, Edgar Martins knows well. It is in the darkness, in the failing light and the black sky that he brings together the social and the metaphysical dimensions that inhabit his work, including this. And in the darkness we can detect the absence of any alternative historical narrative in this moment or at least the inability of America to imagine one. It recalls the nothingness that stands beyond the door in Sartre's *Huis Clos / No Exit*; or around Beckett's 'places', signifying that there is no outside of this.

For Heidegger, dwelling forms part of the grounds of our Being in the world. Dwelling, he writes, is none other than, "the relationship between man and space" (Heidegger, 1975: 157). If dwelling has been undermined then, in the darkness converging on the house we can detect the presence of non-Being. In this, the effect of Martins' visualisations is to complete the circle in which the historical and the metaphysical meet.

5

Reality / Real Estate / Realtors / the 'Real Economy'. Some discomfort has been expressed over Martins' limited digital reshaping of images in this work as if this will have diminished any claims it may have to authenticity. It is a curious reaction given the non-Social Documentary nature of his back catalogue. Furthermore, it appears to reveal a misunderstanding of photography's present situation. This work is themed around a crisis and, as we shall see, it is also work formed by a crisis in its own medium.

Now the word *crisis* is rooted in Greek words meaning 'decision'. The word *decision* has connotations of 'cutting', of 'down cutting'. *This Is Not A House* is a photographic intervention into a crisis. Before it is a set of images, it is a series of decisions, of cuttings into. In a certain sense the work is a set of fictions. And fiction is a way of cutting into, cutting out and rearranging the Real. Ranciére states that, "the real must be fictionalised in order to be thought" (Ranciére, 2006: 38). So, the Real must be transformed (and translated?) in order to be understood. The word *fiction*, has connotations of a false move used to produce real effects (a *feint*), and at the same time it signifies the idea of something made (*fact* and *factory* share its roots), something actual though manufactured (Kermode, 2000: II). Fictions, like decisions, and crises, are made. However much we continue to accept photography's indexical facility, to photograph something is necessarily to fictionalise it, to select, intensify, to link the abstractions (cut) from the visual continuum into sequences and so on are all acts of fictionalising. Photographs establish not so much a reproduction of the Real, as a relationship to that federation of perceptions and reflections, discourses and simulations that, beyond brute materiality, makes up the Real. As the producer of consciously 'aesthetic' work Martins is not attempting to simply record actuality. The work *plays* on the borders between what Ranciére calls, "the logic of facts and the logic of fictions" (Ranciére, 2006: 35); it is itself formed from that relationship, a product of another crisis in which the categories of Art Photography and Documentary Photography have not so much fallen apart as fallen into each other. Martins' work is the bearer of the crisis with which it engages, a crisis that is only in part economic. More immediately the manifest theme of *This Is Not A House* places it at a point where parallel themes converge, each governed by a fear of the loss of reality: the concern that digitalisation has undermined photographic realism and compromised any truth-telling facility in the medium and beyond; and the fear of Wall

Street's undermining of what is interestingly referred to as the *real* economy. The philosopher Peter Osborne [1] has suggested that such anxieties over the epistemological turbulence caused by the ubiquity of digitalisation are themselves the displaced instances of the more fundamental unease experienced over the loss of the real in the Snakes and Ladders practices of finance Capital whose repercussions have spread virally far beyond the obvious domains of the business quarter. Labyrinthine complexities of ownership, the speed-of-light transactions, the fictitious financial products, imaginary capital assets, the hedges and the futures, the toxic debts, foreclosed mortgages, and bankrupted businesses, are the true causes and the real effects of a loss of reality, of a world that can barely be thought. Martins' real fictions though are foregrounded – in clear sight. Wall Street's fictions fooled almost everybody.

6

Martins interferes in a Real that has already been badly interfered with. His interference is made visible most dramatically through the use of the sequence of images of often precariously balanced objects forming assemblages or constructions – I'll call them sculptures – erected by Martins in the deserted interiors from the debris and bric-a-brac left behind. The images are distributed through the work, acting as a kind of *dissonant counterpoint* against the dominant theme of the panoramic or formalised imagery of buildings and landscapes. The sculptures resemble the near accidental found assemblages photographed by Richard Wentworth in the streets of Berlin (Berlin 117), or the playful sculptures of Fischli and Weiss and of the New Zealand artist Paul Cullen. They are like note-taking with objects; a poetic *bricolage*. In the way in which their precariousness mimics that of the buildings all around them, they might be seen as a form of commentary. Maybe they are the expression of Martins' desire to make something out of the uselessness around him, a playful gesture of creativity against the desolation. For sure, there is an attempt here to evolve a 'form of visibility' in which the two contrasting sequences of representation and practice focus our attention onto the overall *madeness* of the work as a whole; that is, its fictiveness. They are markers, the indices of Martins' activity – the photographs of others' constructions and those of his own – as, in both sequences, an arrangement of objects, a manipulation. They record Martins' own performance, his inclusion in the world represented, his responsibility for its depiction. They echo here Jeff Wall's "gestures of reportage and performance" (Stimson, 2005: 109), his "subjectivised witness" (Rosler, 2004: 211). In the light of this, Martins' utilisation of digital processes is underscored as no more or less a manipulation than any of the other processes and strategies he has employed.

7

The real has become unreal. The deserted buildings suck presence out of the landscape. Some interiors, with their uncovered structures and scattered components of heating and ventilating systems, take on the appearance of gallery installations. Like the abstractions of the economic system that has brought them to this state they have become pure forms with no content. Some of the buildings resemble doll's houses or miniature model villages. It is not photography that has made them strange but the economy upon which these houses and interiors were founded – an economy as illusory as the interiors in a Thomas Demand picture.

This Is Not A House produces not a singular truth but a process of truthfully 'recomplicating reality' [2], one embodied in the evidence of its own beauty and founded in the documentation of its own aesthetics.

NOTES

1. Until its scandalous closure in 2010, Peter Osborne, not to be confused with the author of this essay, was Professor of Philosophy in the Centre for Research in Modern European Philosophy at the University of Middlesex. He has accompanied the Centre to its new location at Kingston University in South-West London.

2. A phrase used by Don DeLillo to describe the function of the novel.

REFERENCES

Gaston Bachelard, (1969), *The Poetics of Space*, (Chapter 1, 'House. From Cellar to Garret. The Significance of the Hut'. Translated by Maria Jolas, Boston: Beacon Press.

David Campany, 'Safety in Numbness: Some remarks on the problem of Late Photography', in: Green and Lowry.

David Green and Joanne Lowry (Eds.), (2003), *Where is the Photograph?*, Brighton, Photoworks.

Martin Heidegger, (1975), 'Building, Dwelling, Thinking.' in his, *Poetry, Language, Thought*, translated by Albert Hofstadter, Harper Colophon Books, New York, 1975.

Robin Kelsey and Blake Stimson (Eds.), 2005, *The Meaning of Photography*, New Haven and London, Yale UP.

Frank Kermode, (2000), *The Sense of an Ending, Studies in the Theory of Fiction*, 2nd Edition, Oxford: OUP.

Henri Lefebvre, (1995), *Introduction to Modernity*, London, Verso.

Jean-Luc Nancy, (2005), *The Ground of the Image*, New York, Fordham UP.

Jacques Ranciére, (2006), *The Politics of Aesthetics*, London/NY: Continuum.

Martha Rosler, (2004), *Decoys and Disruptions – selected writings 1975-2001*, Cambridge Mass., London: MIT Press/October Books.

Blake Stimson, 'A Photograph is Never Alone' in: Kelsey and Stimson.

It is true that the idea of objective truth may no longer hold in critical theory, but this work has never been about asserting artistic authorship. Photography is a medium built around conceptual tensions and so it offers me a means to bring together irresolvable contradictions.

This project sought to catalyse and reunite fresh experiences of a new form of American architecture by summoning a disquieting conjunction of realism and fiction.

Fiction – as the essays in this book have hinted at – is elementary to understanding and unravelling the real. Over time we have accepted the fictions we have constructed as facts, making us the forgotten authors of our own narratives.

The houses depicted in this series do not refer just to the particular. They are images of spatial assemblages, of kinds of stages on which a number of quite different (and perhaps incompatible) narratives might be enacted.

These images, these buildings, these ruins, reflect back at us the human constructs that we project and impose on them.

This Is Not A House emerges precisely at that juncture where clear words falter, where language is disturbed. It hurls us into the antinomies of perception and existence, the exploration of limits, the horizon, unstable boundaries. The work traps our visual literacy to the point where we no longer recognise ourselves in the referent. As if nothing were guaranteed.

The meaning of the world is no longer carried on its surface, if indeed it ever was.

Edgar Martins

Peter D. Osborne is a senior lecturer in Cultural Studies and in the History and Theory of Photography in the Faculty of Media at London College of Communication, a constituent college of the University of the Arts London. He is the author of *Travelling Light, photography, travel and visual culture*, Manchester University Press, and numerous articles, papers and catalogue essays on photography and landscape; art, photography and travel; and on Latin American photography. Osborne is presently working on a book titled, *Commemorating the Present – photography and the contemporary cultural condition*. He is on the editorial board of the New Zealand art theory journal, *Z/X*.

Sacha Craddock is an independent art critic and curator. Co-curator for Bloomberg Space and respected post-graduate tutor, she writes numerous articles and catalogue essays, gives public lectures and makes television appearances. Craddock serves in many consultative and official capacities; she is the Chair of New Contemporaries, Chair of Braziers International Workshop, Co-founder of Artschool Palestine and Public Art Advisor for the Royal Borough of Kensington and Chelsea. Recent publications include a section in *Goodbye London, Radical Art and Politics in 1970s London*, recently published in Germany, and a book on Angus Fairhurst. After studying painting at St Martins and then Chelsea School of Art, she started to write criticism in *The Guardian* and then *The Times*. Craddock has judged numerous art prizes, including the Turner Prize in 1999 and the John Moores Painting Prize in 2008.